Ilya Kapovich, Alexei Myasnikov, Vladimir Shpilrain
Combinatorial Group Theory

Also of Interest

Elements of Discrete Mathematics
Numbers and Counting, Groups, Graphs, Orders and Lattices
Volker Diekert, Manfred Kufleitner, Gerhard Rosenberger, Ulrich Hertrampf, 2025
ISBN 978-3-11-106069-9, e-ISBN (PDF) 978-3-11-106255-6

Discrete Mathematics
Combinatorics, Counting, Proofs, Recurrences, Solutions
Walter Hower, 2025
ISBN 978-3-11-120643-1, e-ISBN (PDF) 978-3-11-120689-9

Finitely Presented Groups
With Applications in Post-Quantum Cryptography and Artificial Intelligence
Edited by Volker Diekert, Martin Kreuzer, 2024
ISBN 978-3-11-147337-6, e-ISBN (PDF) 978-3-11-147357-4

Number Theory
Multiplicative and Additive with Factorization and Primality Testing
Peter D. Schumer, 2025
ISBN 978-3-11-157867-5, e-ISBN (PDF) 978-3-11-157928-3

Characters of Groups and Lattices over Orders
From Ordinary to Integral Representation Theory
Alexander Zimmermann, 2022
ISBN 978-3-11-070243-9, e-ISBN (PDF) 978-3-11-070244-6

Abstract Algebra
With Applications to Galois Theory, Algebraic Geometry, Representation Theory and Cryptography
Gerhard Rosenberger, Annika Schürenberg, Leonard Wienke, 2024
ISBN 978-3-11-113951-7, e-ISBN (PDF) 978-3-11-114252-4

Ilya Kapovich, Alexei Myasnikov, Vladimir Shpilrain

Combinatorial Group Theory

Open Problems on Infinite Groups

DE GRUYTER

Mathematics Subject Classification 2020
Primary: 20-02, 20Exx, 20Fxx, 20Jxx; Secondary: 57Mxx, 03D15

Authors

Ilya Kapovich
Department of Mathematics and Statistics
Hunter College
New York
NY 10065
USA
ilya.kapovich@hunter.cuny.edu

Vladimir Shpilrain
Department of Mathematics
The City College of New York
New York
NY 10031
USA
shpilrain@yahoo.com

Alexei Myasnikov
Department of Mathematical Sciences
Stevens Institute of Technology
Hoboken
NJ 07030
USA
amiasnikov@gmail.com

ISBN 978-3-11-914679-1
e-ISBN (PDF) 978-3-11-220813-7
e-ISBN (EPUB) 978-3-11-220838-0

Library of Congress Control Number: 2026930256

Bibliographic information published by the Deutsche Nationalbibliothek
The Deutsche Nationalbibliothek lists this publication in the Deutsche Nationalbibliografie; detailed bibliographic data are available on the Internet at http://dnb.dnb.de.

© 2026 Walter de Gruyter GmbH, Berlin/Boston, Genthiner Straße 13, 10785 Berlin
Cover image: mustafahacalaki / DigitalVision Vectors / Getty Images
Typesetting: VTeX UAB, Lithuania

www.degruyterbrill.com
Questions about General Product Safety Regulation:
productsafety@degruyterbrill.com

Introduction

This is a collection of about 200 open problems in combinatorial group theory.

We are aware of other collections of open problems in group theory including M. Bestvina's (relatively short) list of questions in geometric group theory [45]. In comparison, our problems are more of combinatorial than geometric nature, although we did include several problems with geometric flavor where we felt this was natural.

As a side note, there are two classical monographs, [256] and [264], with the same "Combinatorial group theory" title. One can say that most of our problems are closer in spirit to the "purely combinatorial" monograph [264], whereas the monograph [256] signaled the beginning of a transition to what is now known as geometric group theory that experienced explosive growth one decade thereafter, following M. Gromov's introduction of hyperbolic groups [146]. Then, about 20 years later, a new direction in group theory, at the interface with theoretical computer science, began to emerge; the book [23] gives a good overview of the relevant research avenues. We included in our collection some problems relevant to this emerging area. We gave just a few "sample" problems in that area, mostly on complexity of group-theoretic algorithms.

There is also a well-established collection of open problems titled *The Kourovka notebook: Unsolved problems in group theory* [371] where problems about *finite groups* and groups with various finiteness conditions take a prominent place. We, on the other hand, stay away from finite groups in our collection. Also, unlike [371], we strive to give a background to most problems on our list, except for a few problems whose statements and motivation are "self-explanatory."

In selecting the problems, our choices have been, in part, determined by our own taste. In view of this, we have welcomed suggestions from other members of the community. We want to emphasize that many people have contributed to the present collection, especially to the background part. In particular, we are grateful to G. Bergman, G. Conner, W. Dicks, R. Gilman, V. Remeslennikov, V. Roman'kov, E. Ventura, and D. Wise for useful comments and discussions. We want to single out Gilbert Baumslag, one of the most prominent group theorists of the twentieth century, who passed away in 2014. He had enthusiastically supported this project and contributed several open problems to preliminary versions of this collection [38, 39].

One thing concerning our policy that we would like to point out here is that we have decided to keep on our list those problems that have been solved since we believe those problems are an important part of the list anyway, because of their connections to other, yet unsolved, problems. Solved problems are marked by a $*$, and a reference to a solution, sometimes with a history that led to a solution, is provided in the background. Where there is only a partial solution, we mention that as well.

We have arranged the problems under the following headings:
- Outstanding problems
- Free groups
- One-relator groups

- Finitely presented groups
- Mapping class groups
- Growth
- Equations in and over groups
- Algorithmic problems
- Complexity of algorithms
- Groups of matrices
- Hyperbolic and automatic groups
- Nilpotent groups
- Metabelian groups, and
- Solvable groups

We are aware that nilpotent groups, as well as metabelian groups, are solvable, but they are very special classes of solvable groups, so we think they deserve separate sections.

In the concluding section that we call the "Hall of fame," we have put together a list of names of people who have published solutions of one or more problems from our list.

We want to emphasize that all references we give and attributions we make reflect our personal opinion based on the information we have. In particular, if we are aware that a problem was raised by a specific person, we make mention of that here. We welcome any additional information and/or corrections on these issues.

Contents

Introduction — V

1 Outstanding problems — 1

2 Free groups — 6

3 One-relator groups — 19

4 Finitely presented groups — 24

5 Mapping class groups — 27

6 Growth — 30

7 Equations in and over groups — 32

8 Algorithmic problems — 34

9 Complexity of algorithms — 36

10 Groups of matrices — 39

11 Hyperbolic and automatic groups — 40

12 Nilpotent groups — 45

13 Metabelian groups — 48

14 Solvable groups — 49

15 Overview of recent progress in classical areas — 51
15.1 Recent progress on free groups — 51
15.1.1 Orbit-blocking words — 52
15.1.2 Primitivity-blocking words — 53
15.1.3 Whitehead's automorphism problem — 54
15.1.4 Generic- and average-case complexity of Whitehead's problem — 54
15.1.5 Potentially positive elements — 56
15.1.6 Translation equivalence — 57
15.1.7 Discrete optimization problems — 57
15.2 Recent progress on one-relator groups — 58

15.2.1 Virtually special groups and quasiconvex hierarchies —— **59**
15.2.2 Nonpositive immersions and coherence —— **61**

16   Hall of fame —— 64

Bibliography —— 67

1 Outstanding problems

These are problems that we call "outstanding" because they have attracted a lot of attention from many mathematicians over many years. Other synonyms could be "famous" or "prominent," or "popular", but our first choice was "outstanding."

(O1) (The Andrews–Curtis conjecture). *Let $F = F_n$ be the free group of a finite rank $n \geq 2$ with a fixed set $X = \{x_1, \ldots, x_n\}$ of free generators. A set $Y = \{y_1, \ldots, y_n\}$ of elements of F generates the group F as a normal subgroup if and only if Y is Andrews–Curtis equivalent to X, which means one can get from X to Y by a sequence of Nielsen transformations, together with conjugations by elements of F.*

This problem is of interest in topology as well as in group theory. A topological interpretation of this conjecture was given in the original paper by J. Andrews and M. Curtis [8].

A more interesting topological interpretation arises when one allows one more transformation, namely "stabilization," when Y is extended to $\{y_1, \ldots, y_m, x_v\}$, where x_v is a new free generator (i.e., $y_1, \ldots, y_m$ do not depend on x_v), and inverse of this transformation. Then the Andrews–Curtis conjecture is equivalent to the following (see [389]): two contractible 2-dimensional polyhedra P and Q can both be embedded in a 3-dimensional polyhedron S so that S geometrically contracts to P and Q. Note that this is true if 3 is replaced by 4 – this follows from a result of Whitehead. The problem is amazingly resistant; very few partial results are known. A good group-theoretical survey is [65]. For a topological survey, we refer to [167].

The prevalent opinion is that the conjecture is false; however, not so many potential counterexamples are known. Two of them are given in the survey [65] by R. Burns and O. Macedonska; a one-parameter family of potential counterexamples appears in [4]. More recently, a rather general series of potential counterexamples in rank 2 was reported in [287]. Further potential counterexamples are given in [297].

An interesting result was obtained by S. V. Ivanov [183] who proved that if one replaces "conjugations by elements of F" by just "cyclic permutations" in the statement of the Andrews–Curtis conjecture with "stabilization," one will get an equivalent conjecture.

It might be also of interest that, by using genetic algorithms, A. D. Myasnikov and A. G. Myasnikov [296] showed that all presentations of the trivial group with the total length of relators up to 12 satisfy the Andrews–Curtis conjecture. Thus, the shortest potential counterexample (coming from [4]) is $\langle x, y \mid xyx = yxy, x^3 = y^4 \rangle$.

Finally, we mention a positive solution of a similar problem for *free solvable groups* by A. G. Myasnikov [295].

(O2) (The Burnside problem). *For what values of n are all groups of exponent n locally finite? Of particular interest are $n = 5$, $n = 8$, $n = 9$, and $n = 12$ – the values for which, in experts' opinion, groups of exponent n have a remote chance of being locally finite.*

In contrast with the previous problem (O1), the bibliography on the Burnside problem consists of several hundred papers. We only mention here that E. Golod [135] constructed the first example of a periodic group which is not locally finite; his group, however, does not have a bounded exponent.

The first example of an infinite finitely generated group of a bounded exponent is due to P. S. Novikov and S. I. Adian [310]. We refer to the book [313] for a survey on results up to 1988, and to the papers [179] and [258] for a treatment of the most difficult case where the exponent is a power of 2.

(O3) (Whitehead's asphericity problem). *Is every subcomplex of an aspherical 2-complex aspherical? Or, equivalently, if $G = F/R = \langle x_1, \ldots, x_n \mid r_1, \ldots, r_m, \ldots \rangle$ is an aspherical presentation of a group G (i. e., the corresponding relation module $R/[R,R]$ is a free $\mathbb{Z}G$-module), is every presentation of the form $\langle x_1, \ldots, x_n \mid r_{i_1}, \ldots, r_{i_k}, \ldots \rangle$ aspherical as well?*

This problem received considerable attention in the 1980s. We mention here a paper by J. Huebschmann [173] that contains a wealth of examples of 2-complexes for which Whitehead's asphericity problem has a positive solution. J. Howie [170] points out a connection between Whitehead's problem and some other problems in low-dimensional topology (e. g., the Andrews–Curtis conjecture). We refer to [255] for more bibliography on this problem. Among more recent papers, we mention a paper by E. Luft [253], where he gives a rather elementary self-contained proof of the main result of Howie's paper (see above) and strengthens the result at the same time.

S. V. Ivanov [182] discovered a connection between Whitehead's asphericity problem and Kaplansky's problem (O12)(a) below. A somewhat more general result was later obtained by I. Leary [227].

(O4). *The isomorphism problem for one-relator groups.*

Z. Sela [346] has solved the isomorphism problem for torsion-free hyperbolic groups that do not split (as an amalgamated product or an HNN extension) over the trivial or the infinite cyclic group. It is not known, however, which one-relator groups are hyperbolic (cf. problem (O6)). Earlier partial results are [325] and [327].

We also note that two one-relator groups with relators r_1 and r_2 being isomorphic does not imply that r_1 and r_2 are conjugate by an automorphism of a free group, which deprives one from the most straightforward way of attacking this problem; see [283].

(O5). *The conjugacy problem for one-relator groups.*

The conjugacy problem for one-relator groups with torsion was solved by B. B. Newman [308]. Some other partial results are known; see, e. g., [256] for a survey of older results.

Although many one-relator groups are word-hyperbolic (e. g., for small cancellation reasons or because they have torsion) and hence have solvable conjugacy problem, in full generality this problem remains wide open.

(O6). *Is every finitely generated one-relator group without Baumslag–Solitar subgroups hyperbolic?*

This question is commonly attributed to Gersten and commonly referred to as Gersten's conjecture, although Gersten [127] did not quite pose the question in this form.

Recall that Baumslag–Solitar groups [40] are two-generator one-relator groups with the relator of the form $x^{-1}y^n x = y^m$ for positive $m \neq n$.

Note that every one-relator group with torsion is hyperbolic since the word problem for such a group can be solved by Dehn's algorithm – see the paper [308] by B. B. Newman cited in the background to (O5). Therefore, it suffices to consider torsion-free one-relator groups. We also note that the following weak form of this problem was answered in the affirmative. A group G is called a CSA group if every maximal abelian subgroup M of G is malnormal, i. e., for any element g in G, but not in M, one has $M^g \cap M = \{1\}$.

It is known that every torsion-free hyperbolic group is CSA. Now the following weak form of (O6) holds: A torsion-free one-relator group is CSA if and only if it does not contain metabelian Baumslag–Solitar groups $BS(1,p)$ and subgroups isomorphic to $F_2 \times \mathbb{Z}$ [130].

There has been substantial recent progress related to Gersten's conjecture resulting from the work of Wise, Louder–Wilton [251], and Linton [239].

The work of Helfer–Wise [162] and Louder–Wilton [251] introduced a condition on r, called "negative immersions," which implies that G has no Baumslag–Solitar subgroups. This condition is topological in nature and says that for the presentation complex X of G, for every finite connected complex Y with the Euler characteristic $\chi(Y) \geq 0$ such that Y immerses in X, this Y "Nielsen reduces" to a graph. See [251] for a precise definition.

Louder and Wilton [251] proved that if r has negative immersions, then G has no Baumslag–Solitar subgroups, and they conjectured that, moreover, G is hyperbolic. Recently Linton [239] verified this conjecture and proved that if r has negative immersions then G is indeed hyperbolic. Moreover, Linton showed that in this case G is virtually special and thus residually finite.

Louder and Wilton also provided a more combinatorial/algebraic assumption on r that implies that r has negative immersions. For a nontrivial element $r \in F(X)$, the *primitivity rank* $\pi(r)$ is the smallest rank of a subgroup $H \leq F(X)$ such that $r \in H$ and that r is not primitive in H, if such H exists, and $\pi(w) = \infty$ otherwise. Louder and Wilton proved [251] that if $r \in F(X)$ (where X is finite) and $\pi(r) > 2$, then r has negative immersions. Consequently, by Linton's result, $G = \langle X \mid r = 1 \rangle$ is hyperbolic and residually finite in this case.

We note that the primitivity rank of any given $r \in F(X)$ is algorithmically computable, as was shown in [329].

(O7). *Let G be the direct product of two copies of the free group F_n, $n \geq 2$, generated by $\{x_1, \ldots, x_n\}$ and $\{y_1, \ldots, y_n\}$, respectively. Is it true that every generating system of cardinality $2n$ of the group G is Nielsen equivalent to $\{x_1, \ldots, x_n, y_1, \ldots, y_n\}$?*

The importance of this problem is in its relation to two famous problems in low-dimensional topology, to the Poincaré and Andrews–Curtis conjectures. In particular, R. Craggs [87] showed that the Andrews–Curtis conjecture with "stabilization" (see Problem (O1)b) is true if and only if the answer to Problem (O7) is affirmative. A simplified proof was subsequently given in [88]. See also [144] for more details.

*(O8) (Tarski's problems). *Let $F = F_n$ be the free group of rank n, $Th(F)$ the elementary theory of F, i. e., all sentences in the language of group theory which are true in F.*
(a) *Is it true that $Th(F_2) = Th(F_3)$?*
(b) *Is $Th(F)$ decidable?*

(a) By a result of Yu. Merzlyakov [285], all free groups of finite rank $n \geq 2$ satisfy the same positive sentences.

G. Sacerdote [344] reproved this result, and also proved that all free groups of finite rank $n \geq 2$ satisfy the same ($\forall \exists$) and the same ($\exists \forall$) sentences.

(b) Several fragments of the elementary theory of a free group of finite rank were shown to be decidable. We mention here important work of G. Makanin [266] and A. Razborov [331] on solving equations and systems of equations in a free group. G. Makanin [267] also proved decidability of the universal and positive theories of a free group.

A complete positive solution of both Tarski's problems was announced by O. Kharlampovich and A. Myasnikov in [207]. A subsequent series of papers [205, 208–211] contains a complete exposition of their method with full proofs.

A positive solution of the problem (O8)(a) was also given by Z. Sela in a series of papers [347–353].

*(O9) (The Hanna Neumann conjecture). *If H and K are nontrivial subgroups of a free group, then $rank(H \cap K) - 1 \leq (rank(H) - 1)(rank(K) - 1)$.*

It is convenient to use the notation $(rank - n)(H)$ for $\max(rank(H) - n, 0)$.

Hanna Neumann proved that $(rank-1)(H \cap K) \leq 2(rank-1)(H)(rank-1)(K)$ and conjectured that the coefficient 2 could be removed. R. Burns [64] showed $(rank-1)(H \cap K) \leq (rank-1)(H)(rank-1)(K) + \max((rank-1)(H)(rank-2)(K), (rank-2)(H)(rank-1)(K))$, thus proving the conjectured inequality when both subgroups have rank 2. W. Neumann [306] formulated a stronger version of the Hanna Neumann conjecture, and proved the stronger version of Burns' bound. All subsequent results have applied to the stronger version.

G. Tardos [369] proved the conjectured inequality when one subgroup has rank 2. W. Dicks [97] translated the stronger version into a graph-theoretic conjecture.

G. Tardos [370] improved Burns' bound showing $(rank - 1)(H \cap K) \leq (rank - 1)(H)(rank - 1)(K) + \max((rank - 2)(H)(rank - 2)(K) - 1, 0)$, thus proving the conjectured inequality when both subgroups have rank 3.

W. Dicks and E. Formanek [98] improved Tardos' bound showing $(\text{rank}-1)(H \cap K) \leq (\text{rank}-1)(H)(\text{rank}-1)(K) + (\text{rank}-3)(H)(\text{rank}-3)(K)$, thus proving the conjectured inequality when one of the subgroups has rank 3.

B. Khan [201] showed that if one of the subgroups, say H, has a generating set consisting only of positive words, then H is not part of any counterexample to the conjecture. A similar result was independently obtained by J. Meakin and P. Weil [284].

A complete proof of the conjecture was given by I. Mineyev [288] and independently and almost simultaneously (the proofs were made public within days of each other) by J. Friedman [117].

*(O10). *Is the automorphism group of the free group of rank 2 linear? Or, equivalently, is the braid group B_4 linear?*

E. Formanek and C. Procesi [115] proved that the automorphism group of a free group of rank n is not linear if $n \geq 3$. The "Or, equivalently" statement is due to J. Dyer, E. Formanek, and E. Grossman [106].

The problem was settled in the affirmative by D. Krammer [222]. Later on, S. Bigelow [50] and D. Krammer [223] proved that the Krammer representation of the braid group B_n is faithful for every n, and therefore all braid groups are linear.

(O11). *Is there an infinite finitely presented periodic group?*

Needless to say, infinite finitely generated periodic groups constructed by Golod, Novikov–Adian, and Olshanskii (see the background to (O2)) are infinitely related.

(O12) (I. Kaplansky). *Let G be a torsion-free group, k a field, and kG the group ring of G over k.*
(a) *(I. Kaplansky) Can kG have zero divisors?*
*(b) *Is it true that the group of units of kG is generated by k^* and G?*
(c) *Is it true that the only idempotents of kG are 0 and 1?*

The bibliography on these problems consists of over a hundred papers. One of the highest points here is a result of P. Kropholler, P. Linnell, and J. Moody [225] which implies, in particular, that the integral group ring of a torsion-free virtually solvable group has no zero divisors. We refer to [323] for a survey on results up to 1977.

T. Delzant [96] showed that group rings of a large class of torsion-free hyperbolic groups have no zero divisors.

We also note that S. V. Ivanov [182] discovered a connection between the problem (O12)(a) and Whitehead's asphericity problem (O3).

Part (b) of this problem was recently settled in the negative by G. Gardam [119] for group rings over $\mathbb{F}_2$, the field with two elements. This was later generalized by A. Murray [294] to group rings over fields of arbitrary positive characteristic. Finally, Gardam [120] settled in the negative the characteristic 0 case as well.

2 Free groups

See also problems (O1), (O7), (O8), (O9), and (O10).

In this section, we have collected problems about free groups, their automorphisms, and related issues. Free groups and their properties are at the core of combinatorial group theory, so it is not surprising that this is the largest section in our collection.

Typically, by F_n we denote the free group of a finite rank $n \geq 2$ with a set $X = \{x_1, \ldots, x_n\}$ of free generators. An element $y \in F_n$ is called *primitive* if it is part of a free generating set of F_n or, equivalently, if there is an automorphism of F_n that takes y to x_1.

Most of the basic algorithmic problems about free groups are efficiently solvable, in at most quadratic time. There are classical algorithms, due to Nielsen and Schreier, for solving some less trivial problems, notably the subgroup membership problem. The latter problem was revisited more recently [195, 365] in the context of geometric group theory. The paper [195] also addressed several other algorithmic problems in free groups, including the conjugacy problem for subgroups.

In the twenty-first century, due to emerging interactions of group theory with theoretical computer science, computational complexity of group-theoretic algorithms started to attract considerable attention. In particular, complexity of various classical algorithms was addressed, and often was shown to be lower than previously thought. For example, the complexity of the subgroup membership problem was shown to be quasilinear [376].

One notable exception is Whitehead's automorphism problem: given $u, v \in F_n$, decide whether or not v is an automorphic image of u. The problem is solved by *Whitehead's algorithm* that dates back to 1936 [379]. It is still unknown whether or not the worst-case time complexity of Whitehead's algorithm is bounded by a polynomial function in the length of the input.

Generic- and average-case complexity of Whitehead's problem was studied as well. Notably, the generic-case complexity of Whitehead's algorithm was shown to be linear [197].

(F1). *(a) Is there an algorithm for deciding if a given automorphism of a free group has a nontrivial fixed point?
(b) Is there an algorithm for deciding if a given finitely generated subgroup of a free group is the fixed point group of some automorphism?

S. Gersten [124, 125] proved that the fixed point group $Fix(\varphi)$ of any automorphism φ of a free group F_n of finite rank is finitely generated. A simpler proof was given by D. Cooper [85], and R. Goldstein and E. Turner [134] obtained a similar result for arbitrary endomorphisms of a free group.

M. Bestvina and M. Handel [48] showed that the rank of $Fix(\varphi)$ cannot exceed n. In [178], this was generalized to arbitrary endomorphisms.

All these results, however, do not give an effective procedure for detecting fixed points of a given automorphism. M. Cohen and M. Lustig [78] obtained several useful partial results and, in particular, solved the problem for *positive* automorphisms (i. e., for those that take every free generator to a positive word).

Problem (F1)(a) was settled in the affirmative by O. Bogopolski and O. Maslakova [53, 275] who proved that given an automorphism φ of a free group of finite rank, it is possible to effectively find a (finite) set of generators of the group $Fix(\varphi)$.

For part (b), we note that a subgroup of rank $> n$ cannot possibly be the fixed point group of an automorphism by the result of Bestvina and Handel mentioned above. On the other hand, any cyclic subgroup generated by an element u which is not a proper power is the fixed point group of the inner automorphism induced by u.

A. Martino and E. Ventura [273] gave an explicit description of what an arbitrary fixed subgroup of a finitely generated free group looks like, extending the maximal rank case studied in [82]. However, this description is not a complete characterization, and Problem (F1)(b) therefore remains open.

*(F2) (H. Bass). *Does the automorphism group of a free group satisfy the Tits alternative?*

The Tits alternative for a finitely generated group G is as follows: every finitely generated subgroup of G is either virtually solvable or contains a subgroup isomorphic to the free group on two generators.

Problem (F2) was settled in the affirmative in [47].

*(F3) (V. Shpilrain). *If an endomorphism φ of a free group F_n of finite rank takes every primitive element to another primitive, is φ an automorphism of F?*

This problem was solved in the affirmative for $n = 2$ by V. Shpilrain [359] and by S. Ivanov [181]. S. Ivanov also showed that the answer is positive in the general case under an additional assumption on φ to have a *primitive pair* in the image.

Later, D. Lee [229] settled the problem completely, for every n.

*(F4). *Denote by $Orb_\varphi(u)$ the orbit of an element u of the free group F_n under the action of an automorphism $\varphi \in Aut(F_n)$. That is, $Orb_\varphi(u) = \{v \in F_n, v = \varphi^m(u)$ for some $m \geq 0\}$. If an orbit like that is finite, how many elements can it possibly have if u runs through the whole group F_n, and φ runs through the whole group $Aut(F_n)$?*

There is a nice simple argument showing that the number of elements in an orbit is bounded by a function depending only on n; see [235]. Suppose that for some automorphism φ of $F = F_n$, we have $\varphi^k(g) = g$ and $\varphi^l(g) \neq g$ for $0 < l < k$. Consider the action of φ on the subgroup $H = Fix(\varphi^k)$ consisting of all elements fixed by φ^k. (This subgroup is clearly invariant under φ.) Then φ has order k as an element of $Aut(H)$. Since H has rank at most n by [48], this gives a bound for k in terms of n, since there is a bound for the order of a torsion element in $GL_n(\mathbb{Z})$, hence also for the order of a torsion element in $Aut(F_n)$ because the kernel of the map from $Aut(F_n)$ to $GL_n(\mathbb{Z})$ is torsion-free.

A. Myasnikov and V. Shpilrain [302] showed that the converse is also true, and therefore, in the free group F_n, there is an orbit $Orb_\varphi(u)$ of cardinality k if and only if there is an element of order k in the group $Aut(F_n)$.

Finally, note that possible orders of torsion elements of the group $Aut(F_n)$ were described in [279] and [214]. This completes a solution of Problem (F4).

(F5) (H. Bass). *Is the automorphism group of a free group F_n "rigid," i. e., does it have only finitely many irreducible complex representations in every dimension?*

This problem was settled only for $n=2$ and $n=3$. Specifically, S. Humphries [175] has shown that braid groups are not rigid. This implies that the automorphism group $Aut(F_2)$ is not rigid. Humphries then showed [176] that the group $Aut(F_3)$ is not rigid either.

(F6). *The conjugacy problem for the automorphism group of a free group of finite rank.*

An outer automorphism Φ of a free group F of finite rank is said to be reducible if there is a free factorization $F = F_1 * \cdots * F_k * F'$ such that Φ permutes the conjugacy classes of the subgroups $F_1, \ldots, F_k$; otherwise, Φ is irreducible. Z. Sela [346] and J. Los [248] obtained algorithms that decide if two irreducible outer automorphisms are conjugate in the group of outer automorphisms of F.

Other partial results can be found in [79].

*(F7) (V. Shpilrain). *Denote by $Epi(n, k)$ the set of all homomorphisms from a free group F_n onto a free group F_k; $n, k \geq 2$. Are there elements $g_1, g_2 \in F_n$ with the following property: whenever $\varphi(g_i) = \psi(g_i)$, $i = 1, 2$, for some homomorphisms $\varphi, \psi \in Epi(n, k)$, then $\varphi = \psi$? (In other words, every homomorphism from $Epi(n, k)$ is completely determined by its values on just 2 elements.)*

S. Ivanov [180] proved that every *injective* homomorphism from $Epi(n, k)$ is completely determined by its values on just 2 elements.

Later, D. Lee [228] settled the problem (in the affirmative) completely.

(F8) (W. Dicks, E. Ventura). *Let φ be an endomorphism of a free group F_n and S a subgroup of F_n having finite rank. Is it true that $\text{rank}(Fix(\varphi) \cap S) \leq \text{rank}(S)$?*

This is true if $S = F_n$, although the only known proof of this fact is highly nontrivial (see [48] for the case where φ is an automorphism, and [178] for an extension of this result to arbitrary endomorphisms). For S an arbitrary finite rank subgroup of F_n, the result was established in the case where φ is injective [100].

In [272], it was proved that for an arbitrary family of endomorphisms $\{\varphi_i, i \in I\}$, of the group F_n, the inequality $\text{rank}(\bigcap_{i \in I} Fix(\varphi_i)) \leq \text{rank}(M)$ is true for any subgroup $M \leq F_n$ containing $\bigcap_{i \in I} Fix(\varphi_i)$.

(F9) (A. I. Kostrikin). *Let F be the free group of rank 2 generated by x, y. Is the commutator $[x, y, y, y, y, y, y]$ a product of fifth powers in F? (If not, then the Burnside group $B(2, 5)$ is infinite.)*

The claim in parenthesis requires some justification, so here it is. A. I. Kostrikin has noticed that finite groups of exponent 5 satisfy the Engel identity $[x, y, y, y, y, y, y] = 1$ (see his monograph [220] for more details). Therefore, if the group $B(2, 5)$ were finite, then the element $[x,y,y,y,y,y,y]$ would be a product of fifth powers in F since $B(2,5)$ is the factor group of F over the normal subgroup generated by all products of the fifth powers of elements of F.

*(F10) (A. I. Mal'cev). *Can one describe the commutator subgroup of a free group by a first-order formula in the language of group theory?*

The negative answer to this problem follows from a positive solution of Tarskii's problem (O8)(b) by O. Kharlampovich and A. Myasnikov (see the background to problem (O8)).

Indeed, if the answer to the problem (F10) were positive, this would imply that the elementary theory of a free nonabelian group F (with constants from F in the language) is undecidable, since there is no algorithm for deciding if a given equation in a free group F has solutions from $[F,F]$, according to [105].

In fact, cyclic subgroups are the only definable proper subgroups in free groups and, more generally, in torsion-free hyperbolic groups, see [212].

*(F11) (G. Bergman). *Let S be a subgroup of a free group $F = F_n$ and R a retract of F. Is it true that the intersection of R and S is a retract of S?*

A subgroup H of a group G is termed a retract if there is an endomorphism of the group G that maps G surjectively onto the subgroup H and is identity on H.

First, we note that the intersection of two retracts of a free group is itself a retract, but a proof of this fact is much harder than one would expect, see [44].

I. Snopce, S. Tanushevski, and P. Zalesskii [364] answered this problem completely, as follows. If H is a subgroup of F with rank$(H) = 2$ and R is a retract of F, then $H \cap R$ is a retract of H. On the other hand, for every $m \geq 3$ and every $1 \leq k \leq n-1$, there exist a subgroup H of F_n of rank m and a retract R of F_n of rank k such that $H \cap R$ is not a retract of H.

*(F12) (G. Baumslag). *Let $F = F_n$ be a free group generated by $\{x_1, \ldots, x_n\}$, and let F^Q be the free Q-group, i. e., the free object of rank n in the category of uniquely divisible groups. Consider the map $x_i \longrightarrow (1+x_i)$ from the generators of F^Q into the formal power series ring $Q\langle\langle x_1, \ldots, x_n\rangle\rangle$ with coefficients in Q. It is known that this map induces a homomorphism $\lambda : F^Q \longrightarrow Q\langle\langle x_1, \ldots, x_n\rangle\rangle$ (the Magnus homomorphism). Is λ injective? Or, equivalently, is the group F^Q residually torsion-free nilpotent?*

A construction of the group F^Q in terms of free products with amalgamation is given in [25].

The best known result about the Magnus homomorphism of the group F^Q was due to G. Baumslag [26]. He proved that the Magnus homomorphism is one-to-one on any subgroup of F^Q of the form $\langle F, t | u = t^n\rangle$.

This problem can be reformulated in a more general form, where the ring Q of rationals is replaced by some other associative ring A. In [300], it was shown how to construct a free group F^A for an arbitrary unitary associative ring A of characteristic 0.

In particular, if $A = Z[X]$ is a ring of polynomials with integral coefficients, then F^A is Lyndon's free group.

In [118], it was shown that the Magnus homomorphism of $F^{Z[x]}$ into the corresponding power series ring is an embedding. Moreover, the Magnus homomorphism is an embedding for *every* unitary associative ring A of characteristic 0 if and only if it is an embedding in the case where $A = Q$.

Recently, Jaikin-Zapirain [187] gave a positive answer to Problem (F12).

(F13) (I. Kapovich). *Is the group F^Q in the problem* (F12) *linear?*

We note that Lyndon's free $\mathbb{Z}[x]$-group $F^{Z[x]}$ (see the background to (F12)) is linear. Indeed, the group $F^{Z[x]}$ is discriminated by F [254], hence it is universally equivalent to F, therefore it is embeddable into an ultrapower of F, which is linear.

In [187], in addition to providing a positive solution to Problem (F12), A. Jaikin-Zapirain also proved that the group F^Q is locally linear over $\mathbb{Z}$.

(F14). *Let F be a noncyclic free group of finite rank, and G a finitely generated residually finite group. Is G isomorphic to F if it has the same set of finite homomorphic images as F?*

We note that the answer is "yes" for a *free metabelian* group of finite rank; see [309]. More specifically, if a metabelian group G has the same set of finite homomorphic images as a free metabelian group M of finite rank, then G is isomorphic to M. We note, however, that the paper [309] is still unpublished at the time of this writing.

(F15) (V. Shpilrain). *Let F be a noncyclic free group, and R a noncyclic subgroup of F. Suppose that the commutator subgroup $[R,R]$ is a normal subgroup of F. Is R necessarily a normal subgroup of F?*

This question was motivated by the following result of [14]: if R and S are normal subgroups of F, and $[R,R] \subseteq [S,S]$, then $R \subseteq S$. M. Dunwoody [103] showed that the condition on R being normal cannot be dropped, but it is not known whether or not the condition on S being normal can be dropped.

*(F16) (V. Remeslennikov). *Let R be the normal closure of an element r in a free group F with the natural length function, and suppose that s is an element of minimal length in R. Is it true that s is conjugate to one of the following elements: r, r^{-1}, $[r,f]$, or $[r^{-1},f]$, for some element f?*

This question was motivated by a well-known result of Magnus (see, e.g., [256]): if two elements, r and s, of a free group F have the same normal closure in F, then s is conjugate to r or r^{-1}.

The negative answer was given by J. McCool in [281].

*(F17) (M. Wicks). *Let F_n be a free group of rank $n \geq 2$, and $P(n,k)$ the number of its primitive elements of length k. What is the growth of $P(n,k)$ as a function of k, with n fixed?*

We note that the function $P(n, k)$ is recursive, i.e., its values can be actually computed since there is a simple algorithm for testing whether or not a given element of a free group is primitive.

For $n = 2$, the precise number of *cyclically reduced* primitive elements of length k was given in [302].

It is fairly obvious that $P(n, k) \geq c \cdot (2n - 3)^k$ for some constant c. The problem therefore is to find a tight upper bound for $P(n, k)$.

It was shown in [63] that $P(n, k) \leq c \cdot \mu(n)^k$ for some constant c and some $\mu(n)$ between $2n - 3$ and $2n - 2$, such that $\mu(n) \to (2n - 2)$ as $n \to \infty$.

In [55], it was shown that for some constants c_1, c_2, one has $c_1 \cdot (2n - 3)^k \leq P(n, k) \leq c_2 \cdot (2n - 2)^k$ if $n \geq 3$.

In [361], this was improved to $P(n, k) \leq c \cdot \mu(n)^k$ for $\mu(n)$ strictly less than $2n - 2$, such that $\mu(n) \to (2n - 2)$ as $n \to \infty$.

The problem was completely solved by D. Puder and C. Wu [330]. It was shown that $P(n, k) = (1 + o(1)) \cdot C_n \cdot k(2n - 3)^k$, where C_n is a specific constant that can be computed following the details of the proof.

(F18) (C. Sims). *Is the cth term of the lower central series of a free group of finite rank the normal closure of basic commutators of weight c?*

This is known to be true for $c \leq 5$. For a definition of basic commutators, we refer to [264].

*(F19) (A. Gaglione, D. Spellman). *Let F be a noncyclic free group, and G the Cartesian (unrestricted) product of countably many copies of F. Is the group $G/[G, G]$ torsion-free?*

There is a 2-torsion in this group, see [210].

*(F20) (L. Comerford). *If an equation over a free group F has no solutions in F, is there a finite quotient of F in which the equation has no solutions? (If so, this would provide another proof of Makanin's theorem on solvability of equations in a free group).*

The answer was shown to be negative in [86].

(F21) (P. M. Neumann). *Let G be a free product amalgamating proper subgroups H and K of A and B, respectively. Suppose that A, B, H, K are free groups of finite ranks. Can G be simple?*

It cannot if H and K are of infinite index, see [68]. A. Karrass and D. Solitar [199] proved a result from which one can recover, as a corollary, that the answer is still negative if either of the subgroups H, K has infinite index in A or B, respectively. This corollary was explicitly recovered later by S. V. Ivanov and P. Schupp [186].

M. Burger and S. Mozes [61, 62] settled this problem in the affirmative.

(F22) (A. Olshanskii). *Does the free group of rank 2 have an infinite ascending chain of fully invariant subgroups, each being generated (as a fully invariant subgroup) by a single element?*

It was a major open problem in the theory of varieties of groups (cf. [305]) whether or not every variety can be defined by finitely many identities. This problem was answered in the negative by A. Olshanskii in [312].

It follows that in a free group of infinite rank, there is an infinite ascending chain $V_1 \subset V_2 \subset \cdots$ of fully invariant (a. k. a. verbal) subgroups such that V_1 is generated, as a verbal subgroup, by a single word v_1, V_2 is generated, as a verbal subgroup, by two words v_1, v_2, etc.

Then note that two identities $v_1 = 1$ and $v_2 = 1$ are equivalent to one identity $v_1(X)v_2(Y) = 1$, where X and Y are disjoint sets of variables. Similarly, any verbal subgroup V_k can be generated, as a verbal subgroup, by a single word.

This trick works in a free group of infinite rank but does not work in F_2, the free group of rank 2. This motivated the question.

(F23) (A. Myasnikov, V. Remeslennikov). *Let G be a free product of two isomorphic free groups of finite ranks amalgamated over a finitely generated subgroup.*
(a) Is the conjugacy problem solvable in G?
(b) *Is there an algorithm to decide if G is free?*
(c) *Is there an algorithm to decide if G is hyperbolic?*

If the amalgamated subgroup is cyclic then the first two problems have affirmative answer: (a) is due to S. Lipschutz [241, 242], and (b) is due to Whitehead since a one-relator group is free if and only if the relator is part of a basis of the ambient free group.

It was brought to our attention by C. F. Miller that a slight adjustment of the argument in Theorem 10 of [286] shows that there are free products of two free groups of the same finite rank with finitely generated amalgamated subgroups, that have unsolvable conjugacy problem.

(F24) (O. Kharlampovich, A. Myasnikov). *Is it true that the Diophantine problem in a free group F_r is in the complexity class NP?*

It is known that the Diophantine problem (DP) is NP-hard in free nonabelian groups. It is also known that DP is NP-complete for quadratic equations in free groups [204]. See the discussion in [213].

(F25) (A. Myasnikov, V. Shpilrain). *Let u be an element of a free group F_n, whose length $|u|$ cannot be decreased by any automorphism of F_n. Let $A(u)$ denote the set of elements $\{v \in F_n;\ |v| = |u|,\ \varphi(v) = u \text{ for some } \varphi \in \operatorname{Aut}(F_n)\}$.*
(a) *Is it true that the cardinality of $A(u)$ is bounded by a polynomial function of $|u|$?*
(b) If the free group has rank 2, is it true that the cardinality of $A(u)$ is bounded by $c \cdot |u|^2$ for some constant c, which is independent of u?

This question was motivated by complexity issues for Whitehead's algorithm that determines whether or not a given element of a free group of finite rank is an automorphic image of another given element. It is known that the first part of this algorithm (reducing a given free word to a free word of minimal possible length by "elementary"

Whitehead automorphisms) is pretty fast (of quadratic time with respect to the length of the word). On the other hand, the second part of the algorithm (applied to two words of the same minimal length) was always considered very slow. In fact, the procedure outlined in the original paper by Whitehead suggested this part of the algorithm to be of superexponential time with respect to the length of the words.

However, a standard trick in graph theory shows that there is an algorithm of at most exponential time. Whether or not this algorithm is actually of polynomial time is unknown. The affirmative answer to the problem (F25)(a) would imply that indeed it is.

A. Myasnikov and V. Shpilrain [302] showed that the answer to (F25)(a) is affirmative for the free group of rank 2.

B. Khan [202] showed that a naturally constructed graph of Whitehead transformations in F_2 has very precise geometry. Each connected component is Gromov-hyperbolic as a graph and consists of a small cluster of arbitrary structure with several boundedly fattened infinite tree brunches. Understanding the geometry of this graph confirmed the sharp bound of $8m - 40$ for the cardinality of $A(u)$, $|u| = m$ in F_2 that was suggested by A. Myasnikov and V. Shpilrain based on computer experiments by C. Sims.

D. Lee [230] came close to solving this problem completely. Specifically, she proved that the cardinality of $A(u)$ is bounded by a polynomial function of $|u|$ under the following condition: If two letters x_i (or x_i^{-1}) and x_j (or x_j^{-1}) with $i < j$ occur in u, then the total number of $x_i^{\pm 1}$ occurring in u is strictly less than the total number of $x_j^{\pm 1}$ occurring in u.

Then, D. Lee [231] proved that, under the same assumption on u, the cardinality of $A(u)$ is bounded by a polynomial function of $m = |u|$ of degree $2r - 3$, and that this bound is sharp.

(F26) (M. Bestvina). *Let φ, ψ be two automorphisms of a free group F_n. Is it true that the intersection of $\mathrm{Fix}(\varphi)$ and $\mathrm{Fix}(\psi)$ equals $\mathrm{Fix}(\alpha)$ for some automorphism α of F_n?*

The answer is known to be positive for the free group of rank 2, see [377], and for the free group of rank 3, see [270].

In a free group of arbitrary finite rank, the intersection of $\mathrm{Fix}(\varphi)$ and $\mathrm{Fix}(\psi)$ is always a *free factor* of some $\mathrm{Fix}(\alpha)$, see [271].

(F27) (W. Magnus). *Let u be an element of a free group F_n. An element r in F_n is called a normal root of u if u belongs to the normal closure of r in the group F_n. Can an element u, which does not belong to the commutator subgroup $[F_n, F_n]$, have infinitely many nonconjugate normal roots?*

Magnus himself considered some special cases, see [262]. In particular, he showed that if u is primitive, then, up to conjugacy and inversion, the only normal root of u is u itself, whereas if $u = [x, y]$, then, apart from conjugates of u and its inverse, normal roots of u are just the primitive elements of F_2. Magnus also found the normal roots of $x^p y^p$ whenever p is a prime, and A. Steinberg [367] extended this to finding all roots of $x^p y^q$ whenever p, q are primes.

J. McCool [280] showed that if u is of the form $x^k y^l$, then u has only finitely many normal roots, and those can be found algorithmically. He also gave a description of the set of normal roots of any element of the form $[x^k, y]$.

(F28) (S. Sidki). *Let S be a subgroup of index 2 in the group F_2, and let R be an isomorphic copy of S (in F_2). Denote by f an isomorphism between S and R. Is there necessarily a nontrivial subgroup H in S which is invariant under f?*

For a general setup that motivated this problem, we refer to [304].

***(F29)** (W. Dicks, E. Ventura). *Let H be a subgroup of a free group F_n, and let $r(H)$ denote the rank of H. We call H inert if $r(H \cap K)$ is not bigger than $r(K)$ for any subgroup K of F_n. Is every retract of F_n inert?*

Compare to the problems (F8) and (F11). This problem was solved in the positive in [9].

(F30) (W. Dicks, E. Ventura). *Let H be a subgroup of a free group F_n, and let $r(H)$ denote the rank of H. We call H compressed if $r(H)$ is not bigger than $r(K)$ for any subgroup K containing H. If H is compressed in F_n, is H necessarily inert? (See problem (F29).)*

Clearly, if a subgroup of F_n is inert (see Problem (F29)), then it is compressed in F_n. By the Nielsen–Schreier formula, the two notions coincide for subgroups of finite index in F_n.

Note also that, since every retract of F_n is compressed, the affirmative answer to this problem would imply the affirmative answer to the problem (F29).

(F31) (J. Stallings). *The equalizer of two homomorphisms $\alpha, \beta : F_n \to F_m$ is the group $Eq(\alpha, \beta) = \{x \in F_n : \alpha(x) = \beta(x)\}$. Is it true that if α is injective, then the rank of $Eq(\alpha, \beta)$ is at most n?*

In [365], Stallings notes that there are two homomorphisms from a free group of rank 2 to a free group of rank 1, whose equalizer is not finitely generated.

However, if both $m, n \geq 2$, then the equalizer $Eq(\alpha, \beta)$ is finitely generated – this was proved in [134].

In the case where α is injective and β can be lifted to an injective endomorphism of F_m, the rank of $Eq(\alpha, \beta)$ is indeed bounded by n, see [100].

Finally, we note that G. Bergman showed in [44] that if there is a map $\gamma : F_m \to F_n$ such that $\gamma\alpha$ and $\gamma\beta$ are both the identity on F_n, then $Eq(\alpha, \beta)$ is an intersection of free factors of F_n. In particular, the rank of $Eq(\alpha, \beta)$ is bounded by n in that case.

(F32). (a) *An automorphism of a free group F is called an IA-automorphism if it is identical on the abelianization $F/[F,F]$. Obviously, all IA-automorphisms form a (normal) subgroup $IA(F)$ of the group $Aut(F)$. Is the group $IA(F_n)$ finitely presented for $n > 3$?*

***(b)** (Yu. Merzlyakov) *Is the group $IA(F_n)$ linear for $n > 2$?*

By a result of Magnus, the group $IA(F_n)$ is finitely generated for every n. By a classical result of Nielsen, $IA(F_2)$ is isomorphic to F_2 and is therefore finitely presented (see the discussion after Proposition I.4.5 in [256]).

J. McCool and S. Krstic [282] proved that the group $IA(F_3)$ is *not* finitely presented. The problem remains open for $n > 3$.

Part (b) was settled in the negative by V. Bardakov and R. Mikhailov in [20].

(F33) (A. Myasnikov, V. Shpilrain). *Let F_r be the free group of a finite rank r, with generators $x_1, \ldots, x_r$. An element u of F_r is called* positive *if no x_i occurs in u to a negative exponent. An element u is called* potentially positive *if $\alpha(u)$ is positive for some automorphism α of the group F_r. Finally, u is called* stably potentially positive *if it is potentially positive as an element of F_m for some $m \geq r$.*

(a) *Is the property of being potentially positive algorithmically recognizable?*
*(b) *Are there stably potentially positive elements of F_n that are not potentially positive?*
(c) *Let $P(r, n)$ denote the number of potentially positive elements of length n in F_r. What is the growth of $P(r, n)$ as a function of n, with r fixed?*

Originally, these problems were motivated by results of B. Khan [201] and J. Meakin and P. Weil [284], who established the Hanna Neumann conjecture (see Problem (O9)) in the case where one of the subgroups has a positive generating set. Clearly, the cited result remains valid upon replacing "positive" with "potentially positive" or even with "stably potentially positive."

Eventually, the Hanna Neumann conjecture has been completely settled (see the background to Problem (O9)), but a more serious motivation is provided by results of G. Baumslag [28] and D. Wise [384] who showed that one-relator groups with a positive relator are residually solvable (Baumslag) and residually finite (Wise, with an additional small cancellation condition). Obviously, these results remain valid upon replacing "positive" with "potentially positive."

A. Clark and R. Goldstein [75] settled Problem (F34)(b) in the negative by showing that every stably potentially positive element of F_r is potentially positive.

Goldstein [133] and D. Lee [234] offered (exponential time) algorithms for deciding potential positivity in F_2. Recently, Hyde and O'Connor [217] reported an algorithm with the worst-case time complexity $O(n^2)$. This algorithm has linear generic-case complexity. If inputs are cyclically reduced, then the average-case complexity of this algorithm is constant. This is because, as shown in [217], there are potential positivity-blocking words in F_2, i.e., words that cannot be subwords of any potentially positive word. An example would be $xyx^{-1}y^{-1}x$.

On part (c), we note that the number of positive elements of length n in F_r is r^n. The number of potentially positive elements of length n is bounded from below by the number of elements of length n with only positive occurrences of all but one of the generators. The latter number is $> (\frac{r^2+2r-1}{r+1})^n$.

For $r > 3$, a larger lower bound is provided by the number of primitive elements of length n, which is $O(n(2r - 3)^n)$ by [330].

Recently, [101] obtained a tight estimate for the growth rate in the case $r = 2$ by proving that the number of potentially positive elements of length n in F_2 is $O((\lambda + \epsilon)^n)$ for any $\epsilon > 0$, where $\lambda \approx 2.505$ is the largest root of the polynomial $\lambda^4 - 3\lambda^3 + \lambda^2 + \lambda - 1$. This answers part (c) for $r = 2$.

(F34) (J. Wiegold). *Let R be a characteristic subgroup of a free group $F = F_n$. Can F/R be an infinite simple group?*

If there is no such subgroup, it will follow that $r(S^2) = r(S)$ for any infinite finitely generated simple group S, where $r(S)$ is the rank of S, i.e., the minimum number of generators.

(F35). *Is the group $Out(F_3)$ linear?*

E. Formanek and C. Procesi [115] proved that $Out(F_n)$ is not linear for $n > 3$. On the other hand, $Out(F_2)$ is isomorphic to $GL_2(\mathbb{Z})$ and therefore is linear. Thus, $n = 3$ is the only open case.

(F36) (V. Bardakov). *For any element g of a free group F_n define its primitive length as the minimum k such that g is a product of k primitive elements. Is there an algorithm to determine the primitive length of a given element g of F_n?*

It was shown in [21] that F_n has elements of arbitrarily large primitive length.

We also note that R. Grigorchuk and P. Kurchanov [143] reported an algorithm to determine the length of an element of F_n with respect to the set of all conjugates of elements of a fixed basis of F_n.

(F37) (I. Kapovich, P. Schupp).
(a) *Is there an algorithm which, when given two elements u, v of a free group F_n, decides whether or not the cyclic length of $\varphi(u)$ equals the cyclic length of $\varphi(v)$ for every automorphism φ of the group F_n?*

*(b) *Call elements with the property alluded to in part (a) translation equivalent, to simplify the language. Is it true that whenever g is translation equivalent to h in F_n and $w(x, y) \in F(x, y)$ is arbitrary, one has $w(g, h)$ translation equivalent to $w(h, g)$ in F_n?*

(c) *We say that u is boundedly translation equivalent to v if the ratio of the cyclic lengths of $\varphi(u)$ and $\varphi(v)$ is bounded away from 0 and from ∞ when φ runs through all automorphisms of F_n. Is there an algorithm which, when given two elements of a finitely generated free group, decides whether or not they are boundedly translation equivalent?*

For more details on the motivation, see the paper [193]. Here we just say that we call two elements $u, v \in F_n$ translation equivalent in F_n if for every free and discrete isometric action of F_n on an R-tree X we have $l_X(u) = l_X(v)$, where $l_X(g) = \inf_{x \in X} d(x, gx)$ is the translation length of the action.

In the paper [193] it was shown that u is translation equivalent to v in F_n if and only if the cyclic length of $\varphi(u)$ equals the cyclic length of $\varphi(v)$ for every automorphism φ of F_n.

It is not immediately obvious that there are nontrivial instances of translation equivalence in free groups, but the aforementioned paper provides two different sources of translation equivalence; both methods can be iterated and used to produce arbitrarily large finite collections of distinct conjugacy classes in F_n that are pairwise translation equivalent.

The affirmative answer to part (b) was given by D. Lee in [232]. She also reported an algorithm that decides translation equivalence in F_2, thus giving a partial solution to part (a).

In another paper [233], D. Lee offered an algorithm that decides whether or not two given elements u, v of the free group F_2 are boundedly translation equivalent in F_2, thus giving a partial solution to part (c).

(F38) (O. Bogopolski).
(a) Is there an algorithm which, when given a finitely generated subgroup S of a free group F and an element $g \in F$, decides whether or not there is an automorphism of F that takes g to an element of the subgroup S?
*(b) The following special case of part (a) is especially attractive: given a finitely generated subgroup S of a free group F, find out whether or not S contains a primitive element of F.

The special case described in part (b) is indeed algorithmically solvable as shown in [77].

(F39) (V. Shpilrain).
(a) Let u be an element of a free group F_r. Is it true that there is $v \in F_r$ (depending on u) that cannot be a subword of any cyclically reduced $\varphi(u)$, where φ is an automorphism of F_r?
(b) A special case of interest is where $u = [x_1, x_2]$, where x_1 and x_2 are elements of the same free generating set of F_r.

It is well known that there are many primitivity-blocking words (i.e., words that cannot be subwords of any cyclically reduced primitive element of F_r); see, e.g., [21, 177]. This motivates Problem (F40).

For part (b), we recall a classical result of Nielsen (see, e.g., [264]) saying that in F_2, any cyclically reduced automorphic image of $u = [x_1, x_2]$ is either u or u^{-1}, or a cyclic permutation of one of those, which means there are many words that cannot be a subword of any cyclically reduced $\varphi(u)$ in F_2. However, in F_r with $r > 2$, there is no such facility, so part (b) is open for $r > 2$.

Both parts of this problem were recently answered in the affirmative, first if the ambient free group has rank 2 [177], and then for an arbitrary rank [218].

This implies, in particular, that the following version of Whitehead's automorphism problem has constant average-case time complexity: given a *fixed* $u \in F_r$, decide, on an input $v \in F_r$ of length n, whether or not v is an automorphic image of u.

(F40) (D. Puder, C. Wu). *Suppose u is not a primitive element of F_r. Is it true that the exponential growth rate of the automorphic orbit of u is $\sqrt{2r-1}$?*

D. Puder and C. Wu [330] showed that the exponential growth rate of primitive elements is $2r - 3$ and asked what the growth rate of other automorphic orbits is. Their conjecture is that all other automorphic orbits have the growth rate of $\sqrt{2r-1}$ because this is how the number of conjugates of a given element of F_r grows.

(F41) (Post correspondence problem). *Let F_n and F_m be free groups, and let $\varphi, \psi : F_n \to F_m$ be two homomorphisms. Is there an algorithm to find out whether or not there is $x \in F_n$, $x \neq 1$, such that $\varphi(x) = \psi(x)$?*

There is no such algorithm if free groups are replaced by free monoids, as was established by Post himself [326]. In the context of groups, this problem was first introduced and studied in [298]. See also [74] for variations of this problem for free groups, and [73] for adaptations in other groups.

(F42) (V. Dotsenko). *Let S_n denote the set of all elements of length n in a free group F_r, and B_n the set of all elements of length $\leq n$.*
(a) What is the largest number of elements from S_n that freely generate a subgroup of F_r?
(b) The same question for elements from B_n.

(F43) (V. Remeslennikov). *Let G be a finitely generated residually finite group and $F = F_r$ a free group of rank $r \geq 2$. Is it true that if G and F have isomorphic profinite completions, then G and F are isomorphic?*

(F44). *Let $F_r = F(x_1, \ldots, x_r)$ be the free group of rank $r \geq 2$. What is the worst-case complexity of the primitivity problem for compressed words in F_r? That is, what is the worst-case complexity for deciding, given a compressed word W over $\{x_1^{\pm 1}, \ldots, x_r^{\pm 1}\}$ of size n, whether or not the evaluation $w \in F_r$ of W is a primitive element of F_r? In particular, can this problem be solved in polynomial time in n?*

Compressed words, or *straight line programs*, are a compression tool from computer science that turned out to be quite useful for efficiently solving many group-theoretic decision problems. In particular, in 2008 Schleimer [345] used compressed words to prove that the word problem for $Aut(F_r)$ and $Out(F_r)$ is solvable in polynomial time. Prior to that, the only known solutions required a priori exponential time and exponential space. As noted in our Section 15.1, there is currently an active study of "compressed versions" of various decision problems in group theory, where the inputs are given as compressed words rather than as ordinary words. Note that for the primitivity problem for compressed words in F_r stated above, the evaluation $w \in F_r$ of W has at most exponential length $|w|$ in terms of n. For that reason the primitivity problem for compressed words in F_r has an a priori exponential time upper bound. It is unknown if the true worst-case complexity is lower, even for F_2.

3 One-relator groups

See also problems (O4), (O5), (O6) in Chapter 1.

One-relator groups play a key role in combinatorial and geometric group theory. A *one-relator group* is a group G admitting a presentation $G = \langle X|r \rangle$ where r is a cyclically reduced word in $F(X)$. Most questions about one-relator groups easily reduce to the case where X is finite, and it is usually assumed that one-relator groups are finitely generated. Fundamental groups of compact surfaces are one-relator groups. Thinking about their algorithmic properties led Max Dehn to pose, in his famous 1911 paper [93], the main algorithmic problems for finitely presented groups: the word problem, the conjugacy problem, and the isomorphism problem.

In 1930s Magnus achieved major progress in the study of one-relator groups. In a seminal paper [262], Magnus established the so-called *Freiheitssatz*. It states that if $G = \langle X|r \rangle$ (where r is a nontrivial cyclically reduced word and X is of arbitrary cardinality) and if $Y \subset X$ is a subset omitting some generator $x \in X$ such that x or x^{-1} occurs in r, then the subgroup $A = \langle Y \rangle \leq G$ is freely generated by Y. Such a subgroup Y is called a *Magnus* subgroup of G. In a subsequent paper [263] Magnus proved that every finitely generated one-relator group $G = \langle X|r \rangle$ has solvable word problem. In both of these papers Magnus used a certain iterated amalgamated product decomposition method for representing the structure of one-relator groups, later simplified by Moldavanskii using HNN-extensions (see the book by Lyndon and Schupp for a detailed exposition [256]). This approach remains the main structural tool for studying one-relator groups and is known as the *Magnus–Moldavanskii hierarchy*.

Apart from the surface groups, one-relator groups supply many other key examples in combinatorial and geometric group theory. A famous 1962 paper of Baumslag and Solitar [40] proved that the one-relator group $B(2,3) = \langle a,t|t^{-1}a^2t = a^3 \rangle$ is non-Hopfian and hence not residually finite. Groups $B(p,q) = \langle a,t|t^{-1}a^pt = a^q \rangle$, where $p,q \in \mathbb{Z}, p \neq 0, q \neq 0$, are now called *Baumslag–Solitar* groups. Torus knot groups $\langle x,y|x^p = y^q \rangle$ (where p,q are nonzero coprime integers) provide another important topological family of examples of one-relator groups. A 1968 "spelling theorem" of Newman [308] implies that every finitely generated one-relator group with torsion $G = \langle x_1,\ldots,x_m|r^n \rangle$, where r is a nontrivial cyclically reduced word and $n \geq 2$, is word-hyperbolic. The so-called Baumslag–Gersten group $G = \langle a,t|a^{a^t} = a^2 \rangle$, introduced by Baumslag in 1969 [27], has the Dehn function which grows faster than any fixed iterate of the exponential function, as proved by Gersten in 1992 [128].

Despite Magnus' result on the solvability of the word problem mentioned above, the worst-case complexity of the word-problem for one-relator groups is still unknown, see Problem (OR3). Similarly, the conjugacy problem for one-relator groups (Problem (O5)) remains open.

However, in recent years major progress in the study of one-relator groups took place, and some of the long-standing open problems have been fully resolved, including

(OR1)(a) and (OR1)(c). We give an overview of some of these recent developments in Section 15.2.

(OR1) (G. Baumslag).
*(a) Are all one-relator groups with torsion residually finite?
*(b) Are all one-relator groups with torsion coherent? For a background to this problem, see the survey [31].

A complete positive solution to Problem (OR1)(a) has been announced by Wise in [386], and details have been provided in [387]. Wise's solution to Baumslag's problem comes as a byproduct of his work on the virtually Haken conjecture. Wise proves that all one-relator groups with torsion are "virtually special" which implies their residual finiteness. Earlier partial positive results of Wise on this problem can be found in [384].

Recall that a group G is called *coherent* if all finitely generated subgroups of G are finitely presentable.

Problem (OR1)(b) was posed by G. Baumslag in [30, p. 76]. A complete positive solution was provided by Louder and Wilton [250].

*(c) Are all one-relator groups coherent?

This more general version of Baumslag's coherence question was raised by McCammond and Wise [278], who also attributed it to G. Baumslag. It was recently settled in the positive by A. Jaikin-Zapirain and M. Linton [188].

(OR2). *Is the isomorphism problem solvable for one-relator groups with torsion?*

The isomorphism problem for one-relator groups (Problem (O4) in Chapter 1) remains open, even in the more restricted one-relator groups with torsion context. G. Rosenberger [343] provided a solution of the isomorphism problem for "cyclically pinched" one-relator groups. S. Pride [327] solved the isomorphism problem for two-generated one-relator groups with torsion. Kapovich, Schupp, and Shpilrain [197] showed that the isomorphism problem is solvable in linear time for one-relator groups given by "generic" defining relators.

See also the background to Problem (O4).

(OR3) (A. Myasnikov). *Is the complexity of the word problem for every one-relator group quadratic, i. e., is there for every one-relator group an algorithm solving the word problem in quadratic time with respect to the length of a word? In polynomial time?*

Although Magnus provided a solution of the word problem for one-relator groups in 1932, there is still no known worst-case complexity bound of any kind.

(OR4). *Let $G = \langle X|r \rangle$ be a finitely generated one-relator group. Is it true that for every finitely generated subgroup $H \leq G$, the membership problem for H in G is decidable? Does G have solvable uniform subgroup membership problem?*

Recall that for a finitely presented group $G = \langle X|R\rangle$, the uniform subgroup membership problem asks if there is an algorithm that, given a word $w \in F(X)$ and a finite subset $Y \subseteq F(X)$, decides whether or not w belongs to $\langle Y\rangle$ in G.

Not much is known about Problem (OR4), although the recent proof by Wise that one-relator groups with torsion are coherent suggests that one-relator groups with torsion may indeed have solvable uniform subgroup membership problem. In [278] McCammond and Wise proved that if $G = \langle X|r^n = 1\rangle$ with $n \geq 3|r|$, then G is a locally quasiconvex hyperbolic group, and cosequently G has solvable uniform subgroup membership problem.

In full generality this problem remains open. In some cases a positive solution follows with significant extra work from the theory of word-hyperbolic groups. For example, Martinez-Pedroza and Wise proved [274] that if $G = \langle X|r^m = 1\rangle$, where X is finite, r is a nontrivial cyclically reduced word in X and $m \geq 3|r|$, then G is a locally quasiconvex word-hyperbolic group. It follows that such G has solvable (uniform) subgroup membership problem.

Recently, it was shown in [137] that there are one-relator groups with undecidable *submonoid membership* problem.

(OR5). *Is it true that if the relation module of a group G is cyclic, then G is a one-relator group?*

J. Harlander [158] showed that the answer is "yes" in the case where G is finitely generated and solvable.

(OR6) (G. Baumslag). *Let $H = F/R$ be a one-relator group, where R is the normal closure of an element $r \in F$. Then, let $G = F/S$ be another one-relator group, where S is the normal closure of $s = r^k$ for some integer k. Is G residually finite whenever H is?*

See the survey [31].

(OR7) (G. Baumslag). *Let $G = F/R$ be a one-relator group with the relator from $[F,F]$.*
(a) *Is G hopfian?*
*(b) *Is G residually finite?*
*(c) *Is G automatic?*

A solution of the problems (b) and (c) was communicated to us by A. Olshanskii. In fact, the commutator subgroup $[F,F]$ can be replaced here by *any* noncyclic subgroup of a free group F; the answer will still be negative. It follows from a result of A. Olshanskii [315] that for any m, every noncyclic subgroup H of F contains a subgroup K, which is a free group of rank m, with the following property: for any normal subgroup U of K, the intersection of K and the normal closure of U in F is again U.

To apply this result to our situation, take two elements, x and y, that generate a subgroup $K = F_2$ of $H = [F,F]$ with the property described above. Let r be a Baumslag–Solitar relator built on these two elements; for example, take $r = xy^3x^{-1}y^{-2}$. Let U be the normal closure (in K) of r. Then, from what is said in the previous paragraph, it follows that the normal closure of U in F (call it V) intersects K in U. Therefore, the (one-relator) group F/V contains a subgroup KV/V which is isomorphic to a Baumslag–Solitar group, hence F/V can be neither residually finite nor automatic.

(OR8) (G. Baumslag). *The same as* (OR7), *but for a relator of the form* $[u, v]$.

This problem, as well as (OR7)(a), is motivated by the desire to find a nonhopfian one-relator group which is essentially different from any of the Baumslag–Solitar groups [40].

*(OR9) (D. Moldavanskii). *Are two one-relator groups isomorphic if each of them is a homomorphic image of the other?*

This problem has been settled in the negative by A. Borshev and D. Moldavanskii [56]. To describe a counterexample, denote by $G(l, m, k)$ the group with two generators x and y and one defining relation $y^{-1}x^{-k}yx^ly^{-1}x^ky = x^m$. Then the groups $G(18, 2, 2)$ and $G(18, 2, 6)$ are not isomorphic, but each of them is a homomorphic image of the other.

Other examples can be constructed based on the same ideas.

(OR10). *Is every one-relator group without nonabelian metabelian subgroups automatic?*

Note that hyperbolic groups are automatic and, in particular, an amalgamated product of two free groups with finitely generated subgroups amalgamated is hyperbolic if at least one of the subgroups is malnormal [206].

Furthermore, an amalgamated product of two finitely generated abelian groups is automatic [33].

(OR11) (C. Y. Tang). *Are all one-relator groups with torsion conjugacy separable?*

Recall that a group G is called *conjugacy separable* if for any two nonconjugate elements of G there exists a finite quotient group of G where the images of these two elements remain nonconjugate. Being conjugacy separable implies being residually finite and thus this problem is related to (OR1).

(OR12). *Are all freely indecomposable one-relator groups with torsion cohopfian?*

A group G is called cohopfian if any injective homomorphism of G into itself is an automorphism.

(OR13). (a) *Which finitely generated one-relator groups have all generating sets of minimal cardinality Nielsen equivalent to each other?*
(b) *Which finitely generated one-relator groups have only tame automorphisms (i. e., automorphisms induced by automorphisms of the ambient free group)?*

Two tuples of elements of the same group are called Nielsen equivalent if one of them can be taken to the other by a sequence of *Nielsen transformations*; see, e. g., [256].

For surveys on Nielsen equivalence in groups, we refer to [342] and [155].

(OR14) (G. Baumslag, D. Spellman). *Describe one-relator groups that are discriminated by a free group.*

We note that O. Kharlampovich and A. Myasnikov [205] proved that every finitely generated group which is discriminated by a free group can be obtained from a free group by applying finitely many free constructions of a very particular type.

(OR15). *If G is a one-relator group with the property that every subgroup of finite index is again a one-relator group, and every subgroup of infinite index is free, must G be a surface group?*

*(OR16). *Let $S(n)$ be the orientable surface group of genus n.*
(a) *Are the groups $S(n)$ and $S(m)$ ($m, n \geq 2$) elementarily equivalent (i.e., $Th(S(m)) = Th(S(n))$)?*
(b) *Is $S(m)$ elementarily equivalent to F_{2m}, the free group of rank $2m$?*

The answer to both questions is positive. In fact, all groups $S(n)$ are elementarily equivalent to any free group of rank at least 2, by [211].

4 Finitely presented groups

Although finitely generated free groups and one-relator groups are finitely presented, we believe they deserve separate sections, so you will not find them here.

(FP1). *The triviality problem for groups with a balanced presentation (the number of generators equals the number of relators).*

See also problem (O1) in Chapter 1.

(FP2). *Can a nontrivial finitely presented group be isomorphic to its direct square?*

We note that there are infinitely presented (but finitely generated) groups with this property, see [189]. Moreover, the same author has constructed, for any $n \geq 2$, a (infinitely presented) group G isomorphic to its nth direct power G^n, but nonisomorphic to G^k for any k, $1 < k < n$, see [190].

***(FP3)** (R. Bieri, R. Strebel). Is it true that if the relation module of a group G is finitely generated, then G is finitely presented?*

M. Bestvina and N. Brady gave a negative solution of this problem in [46]. Explicit presentations of their groups have been given by W. Dicks and I. Leary in [99].

(FP4) (J. Stallings). *If a finitely presented group is trivial, is it always possible to replace one of the defining relators by a primitive element without changing the group?*

It is easy to see that the answer is affirmative if the group has rank 2; this is due to the fact that in the free group of rank 2, any element is in the normal closure of a primitive element.

However, if the group has rank 3, then the answer is negative as was shown by S. V. Ivanov [184]. The problem remains open for groups of rank greater than 3.

(FP5) (C. Y. Tang). *Is there a nonfree noncyclic finitely presented group all of whose proper subgroups are free?*

(FP6). *Is every knot group virtually free-by-cyclic?*

For various properties of knot groups, we refer to the book [307].

***(FP7)** (G. Baumslag). Is every finitely generated group discriminated by a free group, finitely presented?*

It is, see [205].

***(FP8)** (G. Baumslag). Is a finitely generated free-by-cyclic group finitely presented?*

It is, see [113].

(FP9) (S. Ivanov). *Is every finitely presented Noetherian group virtually polycyclic?*

A group G is called Noetherian if every strictly ascending chain of subgroups in G is finite.

(FP10) (M. I. Kargapolov). *Is every residually finite Noetherian group virtually polycyclic?*

Compare to the problem (FP9).

***(FP11)** (J. Wiegold). *Is every finitely generated perfect group G (i. e., $[G, G] = G$) the normal closure of a single element?*

This problem was recently answered in the negative by L. Chen and Y. Lodha [71]. The counterexample groups they constructed are free products of left-orderable groups; they are even finitely resented.

***(FP12)** (P. Scott). *Let p, q, r be distinct prime numbers. Is the free product $\mathbb{Z}_p * \mathbb{Z}_q * \mathbb{Z}_r$ the normal closure of a single element?*

No, it is not; see [172].

***(FP13)** (B. Fine). *Let G be an n-generator group. Call a set of elements $\{g_1, \ldots, g_k\}$, $k \le n$, a test set for the group G if, whenever $\varphi(g_i) = g_i$, $i = 1, \ldots, k$, for some endomorphism f of the group G, this φ is actually an automorphism of G. The test rank of G is the minimal cardinality of a test set. Can the test rank be equal to 2 if $n > 2$?*

A single element $g \in G$ is called a *test element* (see [357]) if, whenever $\phi(g) = g$ for some endomorphism φ of the group G, this φ is an automorphism of G. Thus, if G has a test element, the test rank of G is 1. For example, any free group of finite rank has test rank 1. On the other hand, there are groups (for example, free abelian groups of finite rank) whose test rank equals their rank. (Obviously, it cannot be bigger than that.)

E. I. Timoshenko [374] proved that a free metabelian group of rank > 2 has test rank 2.

C. F. Rocca, Jr., and E. Turner [335] showed that for any pair of integers (k, n) with $1 \le k \le n$, there are finitely generated abelian groups of rank n and test rank k.

(FP14) (D. Anosov). *Is there a noncyclic finitely presented group each element of which is a conjugate of some power of a single element?*

(FP15) (V. N. Remeslennikov). *Is every countable abelian group embeddable in the center of some finitely presented group?*

***(FP16)** (R. Hirshon). *Let G be a finitely generated residually finite group, and φ an endomorphism of G. Is it true that $\varphi^{k+1}(G)$ is isomorphic to $\varphi^k(G)$ for some k?*

R. Hirshon himself [165] proved the assertion in the case where $\varphi(G)$ has finite index in G. However, the answer is negative in general [383].

(FP17) (J. Makowsky). *Is there an infinite finitely presented group with finitely many conjugacy classes?*

J. Makowsky [268] pointed out that the affirmative answer to this problem would give an example of a complete finitely axiomatizable theory T which is categorical in uncountable cardinals but not ω-categorical. Subsequently, examples of such theories were given by Peretyat'kin [324] and others (see [166, p. 619]).

We also note that S. Ivanov has constructed examples, for big numbers p, of finitely generated (but infinitely presented) infinite groups of period p with precisely p conjugacy classes. These examples are included as Theorem 41.2 in [313].

*(FP18) (V. Guba). *Is there a finitely generated group, other than $\mathbb{Z}_2$, with exactly 2 conjugacy classes?*

Yes, there is. In fact, D. Osin [320] obtained a much more general result: For any $n \geq 2$, there is an uncountable set of pairwise nonisomorphic finitely generated groups with exactly n conjugacy classes.

(FP19) (E. Zelmanov). *Let F_n be the free group of rank n, and $P_{m,n}$ the subgroup of F_n generated by mth powers of all primitive elements of F_n. (This subgroup is obviously normal in F_n.) Is it true that the factor group $BP(n,m) = F_n/P_{m,n}$ is not residually finite for sufficiently large m?*

(FP20) (R. I. Grigorchuk). *Is it true that every finitely presented group contains either a free 2-generator semigroup, or a nilpotent subgroup of finite index?*

(FP21) (R. I. Grigorchuk). *Let F be Thompson's group $\langle x_0, x_1, \ldots ; x_i x_k x_i^{-1} = x_{k+1}, k > i, k = 1, 2, \ldots \rangle$. Is F amenable?*

See [69] for a survey on various properties of Thompson's group. A good recent survey on the amenability problem for Thompson's group F is [151].

(FP22) (V. Guba). *Let $k[F]$ be the group ring of Thompson's group F (see Problem* (FP21)*) over a field k. Does $k[F]$ satisfy the Ore condition? That is, for any $a, b \in k[F]$, are there nonzero $u, v \in k[F]$ such that $au = bv$?*

If the answer to this problem is negative, then Thompson's group F is not amenable, see [150].

5 Mapping class groups

See also Problem (O10) in Chapter 1.

*(B1). *Are braid groups linear?*

There are two canonical representations of braid groups by matrices over Laurent polynomial rings—the Burau and Gassner representations (the latter is actually a representation of the pure braid P_n group which is a subgroup of finite index in the whole braid group B_n). Both of these representations are faithful for $n = 2, 3$ (a general reference here is [51]).

Problem (B1) was settled in the affirmative by S. Bigelow [50] and D. Krammer [223], who proved that the Krammer representation of the braid group B_n is faithful for every n, and therefore all braid groups are linear. For $n = 4$, see also [222].

We note that it is unknown what the lowest dimension of a faithful representation of the group P_n or B_n is. We also note that it is unknown whether or not B_n is representable by matrices over rational numbers, see Problem (B2). If it were, this would imply, in particular, that the word problem in the group B_n is solvable in quasilinear time, see [316].

(B2). (a) *Is the Gassner representation of the pure braid group P_n faithful for every n?*
(b) *Are braid groups representable by matrices over the field of rationals or an algebraic extension of the field of rationals?*

Despite the problem (B1) being resolved, these problems remains open. In particular, representations offered in S. Bigelow [50] and D. Krammer [223] are *not* over an algebraic extension of the field of rationals. If the answer to part (b) is positive, this will imply, in particular, that the word problem in any braid group has quasilinear time complexity, as compared to the currently known quadratic bound. This follows from a result in [316].

A proof of the Gassner representation being faithful for every n (which implies braid groups being linear) was claimed in [15]. However, there is a controversy around this paper since several people believe they have found essential gaps in the proof (see J. S. Birman's review article 98h:20061 in Math. Reviews). This makes us consider Problem (B2)(a) as open.

(B3). *Is the Burau representation of the braid group B_n faithful for $n = 4$?*

The Burau representation was shown to be nonfaithful for $n \geq 10$ in [292], and then for $n \geq 6$ in [247]. More recently, S. Bigelow [49] has shown that the answer is negative for $n = 5$ as well.

On the other hand, it is known that the Burau representation is faithful for $n = 3$ [265]. Thus, the only case that remains open is $n = 4$.

(B4) (J. Birman). *Let $F = F_n$ be the free group of rank n generated by $a_1, \ldots, a_n$. Is there a solution of the equation $y_1 a_1 y_1^{-1} \cdots y_n a_n y_n^{-1} = a_1 \cdots a_n$ with all y_i from the second commutator subgroup F''?*

The answer is "no" if and only if the Gassner representation of the pure braid group P_n is faithful; cf. Problem (B2).

(B5) (J. Birman). *Give necessary and sufficient conditions for a square matrix over Laurent polynomial ring to be the Burau matrix of some braid.*

For a background, see [51].

(B6) (V. Lin). *Let $n \geq 4$.*
(a) *Does the braid group B_n have a nontrivial noninjective endomorphism?*
*(b) *Is it true that every nontrivial endomorphism of the commutator subgroup $[B_n, B_n]$ is an automorphism?*

For a background and discussion on the problems (B6)–(B8), we refer to a preprint by V. Lin [236]. Here we note that automorphisms of braid groups were described in [107].

Part (b) of this problem was answered in the positive (for $n > 6$) by K. Kordek and D. Margalit [219]. See also the comment on Problem (B8) below.

*(B7)** (V. Lin). *Let $n \geq 4$.*
(a) *Does the braid group B_n have a proper torsion-free nonabelian factor group?*
(b) *Does the commutator subgroup $[B_n, B_n]$ have a proper torsion-free factor group?*

S. Humphries [174] has constructed a representation of the group B_n which is shown to provide torsion-free nonabelian factor groups of B_n, as well as of the commutator subgroup $[B_n, B_n]$ for $n < 7$.

P. Linnell and T. Schick [237] provided a complete solution to Problem (B7) by showing that every braid group B_n is residually torsion-free nilpotent-by-finite, in particular has plenty of nontrivial torsion-free quotients. More specifically, if P_n is the pure braid group inside B_n, and $\gamma_k(P_n)$ its kth lower central series subgroup, then $B_n/\gamma_k(P_n)$ is torsion-free for all sufficiently large k. Therefore, $[B_n, B_n]$ is residually torsion-free nilpotent-by-finite, too, which provides a solution to Problem (B6)(b) as well.

*(B8)** (V. Lin). *Let $n \geq 4$.*
(a) *Is it true that every automorphism of the commutator subgroup $[B_n, B_n]$ can be extended to an automorphism of the whole group B_n?*
(b) *Is it true that every non-trivial endomorphism of $[B_n, B_n]$ can be extended to an endomorphism of B_n?*

S. Orevkov [318] settled Problem (B8)(a) in the affirmative for any $n \geq 4$.

By a more recent result of K. Kordek and D. Margalit [219], any nontrivial endomorphism of the commutator subgroup $[B_n, B_n]$ is, in fact, the restriction of an automorphism of B_n for $n > 6$. This answers both (a) and (b) parts of Problem (B8) for $n > 6$.

(B9) (P. Dehornoy). *We call a braid word w σ-positive (resp. σ-negative) if the generator σ_i with minimal index occurs only with positive (resp. negative) exponents in w. Is it true that, for every n, there exists a constant $c(n)$ such that every n-strand braid word of length N is equivalent to a σ-positive or a σ-negative braid word of length $c(n) \cdot N$ at most?*

(B10) (P. Dehornoy). *Say that a braid is special if it can be obtained from the trivial braid by using iteratively the self-distributive exponentiation $a \wedge b = a\, \mathcal{Q}(b)\, \sigma_1\, \mathcal{Q}(a)^{-1}$, where $\mathcal{Q}$ is the shift endomorphism of B_∞ that maps σ_i to σ_{i+1} for every i. How many special braids are there in B_n?*

For a general background to the problems (B9) and (B10), we refer to the monograph [95].

*(B11) (G. Makanin). *Is it true that in any braid group B_n, $g^k = h^k$ for some $k \neq 0$ implies that g is conjugate to h?*

This is true as shown by J. Gonzalez-Meneses [136].

(B12) (V. Shpilrain). *Let K_n be the kernel of the Burau representation of the braid group B_n, $n > 4$. Is the factor group B_n / K_n torsion free?*

An affirmative answer to this problem would also imply a solution of Problem (B7).

*(B13). *Is there an algorithm with subquadratic worst-case time complexity that would solve the word problem in braid groups?*

There are well-known quadratic-time algorithms for solving the word problem in braid groups, see e. g. [52].

It was shown in [316] that in finitely generated groups of matrices over rationals the worst-case time complexity of the word problem is $O(n \log^2 n)$, where n is the length of the input braid word. It is not known, however, whether or not braid groups can be represented by matrices over rationals, even though braid groups are known to be linear, see the background to Problem (B1).

Recently, Bell and Schleimer [42] reported an algorithm with the worst-case time complexity $O(n \log^3 n)$ that solves the word problem in braid groups, thus answering the question in the affirmative.

6 Growth

(G1) (S. I. Adian). *Is it true that a finitely presented group has either polynomial or exponential growth?*

The point here is that there are examples of groups of intermediate growth (between polynomial and exponential), but all these groups are infinitely presented, see [138–140, 145].

(G2) (R. I. Grigorchuk).
(a) *Is it true that there is a gap between polynomial rate of growth and the rate of growth of the function $e^{\sqrt{n}}$ on the scale of growth of finitely generated groups?*
(b) *Is there a finitely generated group G whose growth function is equivalent to $e^{\sqrt{n}}$?*

*(G3) (M. Gromov, J. Lafontaine, P. Pansu). *Is there a group of exponential, but not uniformly exponential growth?*

A group G has uniformly exponential growth if there is $c > 1$ such that, for any generating system of G, the growth function of G with respect to this generating system is at least $O(c^n)$. For a survey on groups of uniformly exponential growth, see [92].

This problem was settled in the negative for the following classes of groups:
- Hyperbolic groups [221].
- One-relator groups [142].
- Solvable groups [319].
- Linear groups [111].

However, the answer turned out to be affirmative, in general; an example of a group of exponential but not uniformly exponential growth was constructed by J. S. Wilson in [382]. Later, L. Bartholdi [22] gave a similar, but much simpler construction.

*(G4) (R. I. Grigorchuk). *Is it true that every finitely generated infinite simple group has exponential growth?*

The answer is negative. V. Nekrashevych [303] gave examples of finitely generated infinite simple groups of intermediate growth.

(G5) (R. I. Grigorchuk). *Is it true that every hereditary just infinite group (i. e., a residually finite group whose every subgroup of finite index is just infinite) has either polynomial or exponential growth?*

(G6) (R. I. Grigorchuk). *Is there a cancellative semigroup of subexponential growth whose quotient group (if it exists) has exponential growth?*

(G7) (R. I. Grigorchuk). *Let G be a group of subexponential growth, and let $f(n)$ be the number of different elements of length at most n in the group G. Does the limit (as $n \to \infty$) of the ratio $f(n+1)/f(n)$ always exist?*

*(G8). Let $[F_r, F_r]$ be the commutator subgroup of the free group F_r of rank r, and let $f(n)$ be the growth function of $[F_r, F_r]$ in F_r (i.e., the number of different elements of length n in $[F_r, F_r]$). Find the asymptotics of $f(n)$.

R. Sharp [355] showed that the asymptotics of $f(n)$ is $C(2r-1)^n/n^{r/2}$ for some constant C.

(G9). Find the growth rate of the free metabelian group of rank r.

It is easy to show that the number of different elements of length n in the free metabelian group of rank r is bounded from below by the number of self-avoiding walks (SAW) of length n on the lattice $\mathbb{Z}^r$. This follows from a more general observation on interpreting trivial elements of a group $F/[R, R]$ by closed walks on the Cayley graph of the group F/R such that every edge is traversed the same number of times in either direction; see, e.g., [102] or [301] for more details.

It turns out, however, that SAW are quite difficult to count; for general background on SAW, we recommend the monograph [261]. Some progress has been made in counting SAW on $\mathbb{Z}^2$, which led to a reasonable lower bound for the growth rate of the free metabelian group of rank 2. More specifically, in [84], the authors established a lower bound of 2.63815853034. Note that the growth rate of the free metabelian group of rank 2 should be strictly less than 3 (which is the growth rate of the free group of rank 2) since a free metabelian group is not free.

(G10) (R. I. Grigorchuk, I. Pak). Is it true that every group of intermediate growth contains two infinite subgroups A and B which commute elementwise, i.e., $ab = ba$ for any $a \in A$, $b \in B$?

This problem is related to a conjecture of Benjamini and Schramm on percolation on Cayley graphs; see [293] for more details.

(G11) (R. I. Grigorchuk, P. de la Harpe).
(a) Calculate the growth series of Baumslag–Solitar groups $B(p, q)$ with $q > p > 1$.
(b) At least, find the asymptotic rates of growth for $B(p, q)$.

For background, we refer to [141] and [91]. The growth series for the automatic and solvable Baumslag–Solitar groups are rational and can be found in [108] and [81], respectively.

7 Equations in and over groups

See also Problems (F20), (M8).

(E1). *Describe groups over which any equation is solvable.*

(E2) (M. Kervaire, F. Laudenbach). *Let $F_n/R = \langle x_1, \ldots, x_n \mid r_1, \ldots, r_m \rangle$ be a presentation of a nontrivial group. Is it true that the group $\langle x_1, \ldots, x_n, x_{n+1} \mid r_1, \ldots, r_m, s \rangle$ is also nontrivial for any element s from F_{n+1}?*

It is clear from considering abelianization that, if $G = F_n/R$ is a counterexample, then G must be perfect, i. e., $G = [G, G]$. Also, it suffices to consider the case where G is an infinite simple group.

There are several related problems about (systems of) equations over groups. We only give one of them here; it appears as Problem 2a on Lyndon's list [255]:

> If, in the above notation, the sum of exponents on x_{n+1} in s is not 0, does the equation $s = 1$ always have a solution over G?

For results on the latter problem, we refer to [215] and [76]. For other related problems, we refer to [171].

(E3) (J. Birman). *Let $F = F_n$ be the free group of rank n generated by $a_1, \ldots, a_n$. Is there a solution of the equation $y_1 a_1 y_1^{-1} \cdots y_n a_n y_n^{-1} = a_1 \cdots a_n$ with all y_i from the second commutator subgroup F''?*

The answer is "no" if and only if the Gassner representation of the pure braid group P_n is faithful; cf. Problem (B2).

(E4) (G. Baumslag, A. Miasnikov, V. Remeslennikov). *Is a free product of two equationally Noetherian groups equationally Noetherian?*

A group is called equationally Noetherian if every system of equations in finitely many unknowns in this group is equivalent to a finite subsystem.

R. Bryant [60] and V. Guba [149] proved that free groups are equationally Noetherian; see also [366].

Theorem 9.1 in the (still unpublished) preprint [354] claims a positive answer to this problem.

For a general discussion on this and related problems, we refer to [37].

(E5) (G. Baumslag, A. Miasnikov, V. Remeslennikov). *Is a free pro-p group equationally Noetherian?*

(E6) (A. Garreta, A. Myasnikov, D. Ovchinnikov).
(a) *Is it true that the Diophantine problem (DP) in every nonvirtually abelian polycyclic group is undecidable?*

https://doi.org/10.1515/9783112208137-007

(b) *Is the Diophantine problem undecidable in the polycyclic group $G = A \rtimes B$, where A is a finitely generated free $\mathcal{O}$-module and B is an infinite subgroup of $\mathcal{O}^*$ for some ring of algebraic integers $\mathcal{O}$?*

It is known that the Diophantine problem is undecidable in every nonvirtually abelian finitely generated *metabelian* polycyclic group [121].

(E7) (A. Garreta, A. Myasnikov, D. Ovchinnikov).
(a) *Is it true that the Diophantine problem in every nonvirtually abelian finitely generated metabelian group is undecidable?*
(b) *Is the Diophantine problem undecidable in metabelian nonabelian Baumslag–Solitar groups $BS(1, n)$?*

We note that Mandel and Ushakov [269] studied quadratic equations in metabelian Baumslag–Solitar groups $BS(1, n)$ and showed, in particular, that the corresponding Diophantine problem is NP-complete whenever $|n| \neq 1$ and decidable in polynomial time otherwise.

(E8) (Yu. Matiyasevich). *Is the problem of solving equations with length constraints decidable in nonabelian free groups or free semigroups?*

"With length constraints" means that the lengths of solutions should satisfy a given system of linear equations with integer coefficients. We refer to [277] for further discussion on this problem.

8 Algorithmic problems

See also Problems (O4), (O5), (O8), (F1), (F6), (OR4), (FP1), (H2), (H3), (M1), (S2), (S5), and (S6).

(A1). *Are there finitely generated 2-relator groups with unsolvable word problem?*

It is a classical result of W. Boone [54] that there are finitely presented groups with unsolvable word problem. The present problem asks how "small" a presentation of a group with unsolvable word problem can be.

We note that finitely generated one-relator groups have solvable word problem; this is another classical result, due to W. Magnus [262].

***(A2)** (D. Collins). *Can every torsion-free group with solvable word problem be embedded in a group with solvable conjugacy problem?*

The answer is negative. Counterexamples were reported by A. Darbinyan [90].

(A3). *Let G be a finitely presented group, H the intersection of all normal subgroups of finite index in G. Can G/H have unsolvable word problem?*

(A4) (L. P. Neuwirth). *Is there an (algebraic) algorithm to decide if a knot group is cyclic?*

We just note here that the importance of this problem is enhanced by a well-known fact that the knot group of a knot K is (infinite) cyclic if and only if K is isotopic to the unknot. See also [381] for a generalization of this fact implying, in particular, that the knot group of a prime knot determines the knot type up to the mirror reflection.

(A5) (V. Shpilrain). *Is the set of finitely presented metabelian groups recursively enumerable?*

All algorithmic decision problems naturally split into "yes" and "no" parts. The "yes" part is usually straightforward in the sense that there is usually a recursive procedure that enumerates all objects for which the answer to the given decision problem is "yes." This is the case, in particular, with the word, as well as the conjugacy problem for a given finitely presented group and with the problem of isomorphism to a given finitely presented group. The situation becomes more complex if one considers the problem of isomorphism to a group from a given class of groups. In this case, one may encounter a situation where neither finitely presented groups in the given class nor finitely presented groups outside of the given class are recursively enumerable. (The desire to have a class like that is motivated by applications to cryptography.) The class of metabelian groups seems to be the simplest candidate, although the same question can be asked about any class of groups that contains infinitely presented finitely generated groups.

(A6) (M. Chiodo). *Is there a general procedure to produce a nontrivial element from a finite presentation of a nontrivial group?*

This problem is discussed in [72], where a special case is settled; namely, it is shown that there is no general procedure to pick a nontrivial generator from a finite presentation of a nontrivial group.

9 Complexity of algorithms

Sometimes, when an algorithmic decision problem is known to be solvable, the focus shifts to finding an efficient (e. g., polynomial-time) algorithm. In this section, we have collected some especially interesting (in our opinion) problems of this kind. Most of them are motivated by (more or less) practical applications. See also Problems (F25), (OR3), and (H4).

*(C1). *Is there a polynomial-time algorithm for solving the word problem in the group $Aut(F_n)$ (with respect to some particular finite presentation), where F_n is the free group of rank n?*

The word problem in $Aut(F_n)$ has a straightforward solution (by just acting on the generators of F_n), but this algorithm is exponential with respect to the length of the input. (The input here is an automorphism given as an n-tuple of words in the generators of $Aut(F_n)$.) S. Schleimer [345] has settled this problem in the affirmative by using the straight line programs technique. See more about this technique in [23].

(C2) (S. Ivanov). *Let u, v be two elements of the free group F_n. According to Whitehead, the problem of detecting whether or not u is an automorphic image of v is decidable. Is this problem in the class NP (that is, decidable in nondeterministic polynomial time with respect to the maximum of $|u|, |v|$)?*

We note that the affirmative answer to Problem (F25)(a) would imply the affirmative answer to this problem, but the converse is not necessarily true. A positive answer to Problem (C2) would basically mean that, given the information that u is an automorphic image of v, one can verify this in time polynomial in $\max(|u|, |v|)$.

See the background to Problem (F25) or Section 15.1.3 for detailed information on what is currently known about complexity of Whitehead's automorphism problem.

*(C3) (V. Shpilrain). *Is the conjugacy problem in the braid group B_n in the class NP (that is, decidable in nondeterministic polynomial time with respect to the maximum of the lengths $|u|, |v|$, where u, v are the input words)?*

Since the first solution of the conjugacy problem in braid groups was given by Garside [122], several other people have come up with different algorithms, including [52, 116, 123], but none of the algorithms suggested so far has polynomial time complexity with respect to the lengths of the input words.

Being in the class NP is a weaker property, i. e., it might be the case that the conjugacy problem in braid groups has no polynomial time solution, but it is in the class NP. The latter would follow, in particular, from the following: given two conjugate elements of B_n represented by braid words of lengths $\leq m$, there is a conjugator whose length is bounded by a polynomial function of m.

We note that M. Calvez and B. Wiest [67] showed that for the braid group B_4, the conjugacy problem is actually in the class P. In fact, it has a cubic-time solution with respect to the lengths of the input words.

This problem was eventually completely settled for all mapping class groups. First, H. A. Masur and Y. N. Minsky [276] obtained a linear upper bound for pairs of conjugate pseudo-Anosov elements in any mapping class group. J. Tao [368] completed the remaining cases. J. Behrstock and C. Drutu [41] gave alternative proofs of these bounds.

(C4). *Is there a subexponential upper bound for the time complexity of Dehornoy's algorithm for solving the word problem in braid groups? (cf. Problem (B9))*

A description of Dehornoy's algorithm can be found in [94]. We note that there is a quadratic time algorithm for solving the word problem in braid groups, which is based on the fact that braid groups are automatic. Dehornoy's algorithm seems to outperform, in real-life implementations, this as well as other known algorithms for solving the word problem in braid groups. However, no subexponential upper bound for the time complexity of Dehornoy's algorithm has been established theoretically.

(C5). *What is the time complexity of the word problem in any finitely generated metabelian group? Is it true that it is at most quadratic?*

We note that the word problem in any finitely presented metabelian group is solvable. This follows from the fact that these groups are residually finite, but the first explicit algorithm was offered by E. I. Timoshenko in [372].

(C6) (A. Olshanskii, V. Shpilrain).
(a) Is there a linear-time algorithm for solving the word problem in finitely generated groups of matrices over integers?
***(b)** Is there a linear-time algorithm for solving the word problem in finitely generated groups of upper unitriangular matrices over integers (in particular, in finitely generated torsion-free nilpotent groups)?

In [316], the authors offered a quasilinear-time algorithm for part (a) of this problem. More specifically, if an input word w in the given generators has length n, then there is an algorithm that tells whether or not w represents the identity matrix and takes time $O(n \log^2 n)$. This result is heavily based on the following result from [159]: there is an algorithm for multiplying two n-bit integers of time complexity at most $C \cdot n \log n$, where C is some positive constant. Since addition of two n-bit integers has complexity $O(n)$, this result implies that multiplication of matrices over $\mathbb{Z}$ whose entries have bit length bounded by m has complexity bounded by $C_1 \cdot m \log m$ for some $C_1 > 0$.

It is unlikely that the result of [159] can be improved, but perhaps there is another approach to the word problem in groups of matrices that could give a tighter complexity estimate.

For part (b), the authors of [316] offered an algorithm that has time complexity $O(n \cdot \log^{(k)} n)$ for any integer $k \geq 1$, where $\log^{(k)} n$ denotes the function $\log(\cdots(\log n))$, with k logarithms.

Finally, part (b) was answered in the positive in [24].

We also note that if there is a linear-time algorithm for the word problem in torsion-free nilpotent groups, then there is also a linear-time algorithm for the word problem in any finitely generated nilpotent groups, with or without torsion, see [316].

10 Groups of matrices

(MA1). *Is the group $GL_2(\mathbb{Z}[t, t^{-1}])$, i.e., the group of invertible matrices over the ring of one-variable Laurent polynomials with integral coefficients, generated by elementary and diagonal matrices?*

See [16] for a general background. In particular, one-variable case is the only case still open (see the next problem).

*(MA2). *Find a particular matrix from $GL_2(\mathbb{Z}[t, t^{-1}, s, s^{-1}])$, which is not a product of elementary and diagonal matrices.*

M. Evans [112] found such matrices. It was previously known due to [16] that such matrices do exist.

(MA3). *The subgroup membership problem for the group $SL_3(\mathbb{Z})$.*

We note that in $SL_4(\mathbb{Z})$, the subgroup membership problem is unsolvable since a direct product of two free groups embeds into $SL_4(\mathbb{Z})$. In $SL_2(\mathbb{Z})$, the subgroup membership problem is long known to be solvable; see, e.g., [246] for a polynomial-time solution.

(MA4) (S. Thomas). *Does there exist a simple torsion-free linear group?*

11 Hyperbolic and automatic groups

The notion of a *word-hyperbolic*, or just *hyperbolic*, group was introduced by Gromov in a seminal 1987 monograph [146] with the goal of generalizing algebraic properties of fundamental groups of compact hyperbolic manifolds and, more generally, capturing the idea of a "large scale" negative curvature for a finitely generated group. Small cancellation theory, particularly as developed by Lyndon and Schupp [256], provided another important precursor for hyperbolic groups.

A finitely generated group G is called *word-hyperbolic* if for some (equivalently, every) finite generating set X of G there exists $\delta \geq 0$ such that all the geodesic triangles in the Cayley graph $Cay(G,X)$ are δ-thin, that is, each such a triangle is contained in the union of the δ-neighborhoods of the other two sides. Word-hyperbolic groups are finitely presentable, and one of equivalent characterizations of hyperbolicity is that a finite group presentation satisfies a linear isoperimetric inequality, see [7]. There are many natural sources of word-hyperbolic groups coming from different parts of mathematics. Fundamental groups of compact strictly negatively curved manifolds are word-hyperbolic, as are groups admitting properly discontinuous isometric actions on proper geodesic δ-hyperbolic metric spaces (that is, with δ-thin geodesic triangles in the above sense). Finite rank free groups and finitely presented $C'(1/6)$ and $C'(1/4) - T(4)$ groups are word-hyperbolic. "Generic" or "random" finitely presented groups in various senses are word-hyperbolic as well [11, 311, 317].

Hyperbolic groups have many good algebraic, algorithmic, geometric, dynamical, and other properties. In particular, of the classic Dehn algorithmic problems, the word problem for hyperbolic groups is solvable in linear time and the conjugacy problem is solvable in $O(n \log n)$ time [110]. Moreover, in 1990s Sela solved the isomorphism problem for torsion-free word-hyperbolic groups [346], and in 2011 Dahmani and Guiradel extended the isomorphism problem solution to all word-hyperbolic groups [89].

Crucially, the property of being word-hyperbolic is a quasiisometry invariant for finitely generated groups [7], which led Gromov to initiate a program [147] of studying which other group-theoretic properties are preserved by quasiisometries.

Since their introduction by Gromov, a deep and rich theory of hyperbolic groups, with many applications and generalizations, has been developed. Every hyperbolic group G can be naturally compactified by adding its *hyperbolic boundary* ∂G, which in turn carries additional useful dynamical and analytic structures. These structures, and their interaction with the geometry of G, have been used to obtain a variety of structural, rigidity, cohomological, etc., results about hyperbolic groups and their specific subclasses, see [192]. Hyperbolic groups also provided the right setting for initiating substantial progress in the study of important K-theoretic conjectures coming from topology, such as the Novikov conjecture and the Baum–Connes conjecture, see [336]. Hyperbolic groups proved well suited for generalizing classic small cancellation theory and constructing quotient groups (often infinitely presented ones) with various prescribed properties, see [314]. "Random" groups in Gromov's density model, near the density parameter $d = 1/2$ (as well as their quotients), provided a rich new source of groups with Kazhdan property (T), see [13].

Several important generalizations of hyperbolic groups have been developed since then, including, in particular, the notions of relatively hyperbolic, acylindrically hyperbolic, and hierarchically hyperbolic groups.

Almost contemporaneously with the introduction by Gromov of the notion of word-hyperbolic groups, the concept of automatic groups arouse in late 1980s and early 1990s [109]. A finitely generated group G is *automatic* if for some (equivalently, every) finite generating set X of G, the group G admits a set of normal forms L given by a regular language over $X^{\pm 1}$ such that L satisfies the so-called "synchronous fellow traveller property" [33]. Word-hyperbolic groups are automatic, but the class of automatic groups is bigger and includes many other interesting examples, such as fundamental groups of complete noncompact hyperbolic manifolds of finite volume, mapping class groups of compact surfaces, etc. Automatic groups are finitely presentable, have word problem solvable in quadratic time, and satisfy a quadratic isoperimetric inequality. Biautomatic groups, a natural subclass of automatic groups, also have solvable conjugacy problem. Compared to hyperbolic groups, the intrinsic theory of automatic groups is much more limited, see [332] for a recent summary.

Both for automatic groups and especially for hyperbolic groups and their generalizations, many significant open problems remain unresolved.

(H1). (a) *Are all hyperbolic groups residually finite?*
(b) *Does every nontrivial hyperbolic group have a proper subgroup of finite index?*

I. Kapovich and D. Wise [198] proved the equivalence of (a) and (b).

*(H2). *Are all hyperbolic groups linear?*

M. Kapovich [191] provided a negative answer to this question. Using "small cancellation" quotients of certain superrigid hyperbolic lattices, he constructed an infinite word-hyperbolic group G such that any finite-dimensional linear representation of G (over any field) has finite image. Therefore G is not linear.

(H3). *Do hyperbolic groups with torsion have solvable isomorphism problem?*

Z. Sela [346] has solved the isomorphism problem for torsion-free hyperbolic groups that do not split (as an amalgamated product or an HNN extension) over the trivial or the infinite cyclic group.

In 2011 F. Dahmani and V. Guirardel [89] gave a solution for the isomorphism problem for all word-hyperbolic groups.

(H4) (A. Myasnikov). *Given a finite presentation of a hyperbolic group (which is not necessarily a Dehn presentation), is it possible to find a Dehn presentation for this group in polynomial time?*

Note that every hyperbolic group has a Dehn presentation, see [257].

(H5) (A. Myasnikov). *Given a finite presentation of an automatic group, can one decide if this group is hyperbolic?*

P. Papasoglu [321] gave a partial algorithm to recognize hyperbolic groups. Given a finite presentation $\langle S, R \rangle$, the algorithm terminates if the group $G = \langle S, R \rangle$ is hyperbolic and gives an estimate of the hyperbolicity constant δ.

(H6) (S. Gersten). *Are all automatic groups biautomatic?*

For a background on problems (H6) through (H10), we refer to [127].

(H7) (S. Gersten). *Does every automatic group have a solvable conjugacy problem?*

(H8) (S. Gersten). *Is every biautomatic group that does not contain any $\mathbb{Z} \times \mathbb{Z}$ subgroups hyperbolic?*

The only currently known source of finitely presented but nonhyperbolic subgroups of hyperbolic groups comes from a construction of N. Brady [57]. Brady constructed a hyperbolic group G and a finitely presented subgroup H of G such that H does not have the homological finiteness property FP_3 and hence is not hyperbolic. The group H has no $\mathbb{Z} \times \mathbb{Z}$ subgroups since H is a subgroup of a hyperbolic group. However, a result of Alonso [6] implies that every automatic group is FP_∞, and therefore in Brady's example H is not automatic.

V. Roman'kov [340] showed that every solvable biautomatic group without $\mathbb{Z} \times \mathbb{Z}$ subgroups is virtually abelian and therefore hyperbolic.

(H9) (S. Gersten). *Can the group $\langle x, y; yxy^{-1} = x^2 \rangle$ be a subgroup of an automatic group?*

(H10) (S. Gersten). *Is a retract of an automatic group automatic?*

(H11). *Does every hyperbolic group act properly discontinuously and cocompactly by isometries on a $CAT(k)$ space, where $k < 0$?*

A key task in geometric group theory is to explore to what extent various coarse large-scale group-theoretic notions are more general than their continuous and local counterparts. When introducing hyperbolic groups in [146], Gromov observed that if a group G admits a properly discontinuous cocompact isometric action on a $CAT(k)$ space with $k < 0$ then G is hyperbolic. He also provided a variety of constructions for showing that various classes of hyperbolic groups do admit such actions, and there has been much subsequent work by other authors in the same direction. However, it is still unknown if the converse to Gromov's result holds [58].

(H12) (G. Baumslag, A. Myasnikov, V. Remeslennikov). *Is every hyperbolic group equationally Noetherian? (A group is called equationally Noetherian if every system of equations in finitely many variables in this group is equivalent to a finite subsystem).*

For a background on equationally Noetherian groups, we refer to [37]. Here we just mention that free groups are equationally Noetherian; this is due to R. Bryant [60] and V. S. Guba [149].

For a general discussion on this and related problems, we refer to [36].

(H13). *Are all combable groups automatic?*

See [109] for a precise definition of being combable. Roughly, a finitely generated group G is *combable* if for some (equivalently, every) finite generating set A of G there exists a collection L of normal forms over $A^{\pm 1}$ for elements of G such that L consists of quasigeodesic words and such that L satisfies the so-called "fellow traveller property." In this case, L is called a *combing* on G.

As shown in [109], for an automatic group G any automatic language for G is a combing. Hence all automatic groups are combable. However, it is still unknown if the converse statement holds.

*(H14) (A. Myasnikov). *A subgroup H of a group G is called malnormal if for every element g in G, such that $g \notin H$, one has $g^{-1}Hg \cap H = \{1\}$. Is this property algorithmically decidable for finitely generated subgroups of a hyperbolic group?*

We note that malnormality is decidable in *free* groups, see [35]. Another idea on how to check malnormality is essentially contained in [365].

However, there is no algorithm that determines for *any* hyperbolic group G and its arbitrary finitely generated subgroup H whether H is malnormal in G or not, see [59].

We note that there exists an algorithm (due to D. Holt) that decides whether or not a given finitely generated *quasiconvex* subgroup of a hyperbolic group is malnormal.

(H15) (A. Myasnikov). *If a finitely generated subgroup H of a hyperbolic group is malnormal (see above), does it follow that H is quasiconvex?*

A subgroup H of a group G has *finite width* in G if there exists an integer $n \geq 1$, such that for any collection $\{Hg_i \mid i \in I\}$ of pairwise distinct cosets of H in G with the property that the intersection $g_i^{-1}Hg_i \cap g_j^{-1}Hg_j$ for all $i \neq j$ in I, the set I has cardinality at most n. Note that a malnormal subgroup H of G satisfies this property with $n = 1$.

A result of R. Gitik, M. Mitra, E. Rips, and M. Sageev [131] shows that if H is a quasiconvex subgroup of a hyperbolic group G then H has finite width in G. Moreover, all known sources of constructing finitely generated nonquasiconvex subgroups of hyperbolic groups produce subgroups of infinite width. Thus a more general version of (H15) would ask if it is true that every finitely generated subgroup H of finite width in a hyperbolic group G is quasiconvex in G.

(H16). *Is every metabelian automatic group virtually abelian?*

We note that a finitely generated nilpotent group is automatic if and only if it is virtually abelian, see [109].

(H17) (A. Ol'shanskii). *Does every hyperbolic group G have a free normal subgroup F such that the quotient group G/F is of finite exponent?*

By a result of S. Ivanov and A. Ol'shanskii [185], every nonelementary hyperbolic group G admits an infinite quotient group $\overline{G}$ of finite exponent.

(H18) (T. Riley). *Can a word-hyperbolic group G contain a finitely presented subgroup H such that the membership problem for H in G is undecidable?*

A famous construction of Rips [334] can be used to construct a small cancellation (and thus word-hyperbolic) group G with a finitely generated subgroup H such that the membership problem for H in G is undecidable. However, the subgroup H in Rips' construction is necessarily not finitely presentable.

(H19) (S. Gersten). *Let F be Thompson's group $\langle x_0, x_1, \ldots; x_i x_k x_i^{-1} = x_{k+1}, k > i, k = 1, 2, \ldots \rangle$. Is F automatic?*

See, e. g., [69] for a survey on various properties of Thompson's group.

(H20) (Cannon's conjecture). *Let G be a word-hyperbolic group such that the boundary of G is homeomorphic to the 2-sphere. Then G admits a C discrete cocompact isometric action on $\mathbb{H}^3$.*

See [192] for the background information on this problem.

(H21) (M. Gromov). *If G is a word-hyperbolic group that is not virtually free, does G contain a surface subgroup?*

This question is commonly attributed to Gromov and sometimes formulated as a conjecture. Various positive partial results are available. For example, Calegari and Walker [66] obtained a positive answer to problem (H21) for "random" finitely presented groups (in Gromov's density model), which are known to be word-hyperbolic.

*(H22) (M. Mitra). *If H is a word-hyperbolic subgroup of a word-hyperbolic group G, does the inclusion $i : H \to G$ always extend to a continuous H-equivariant map $\hat{i} : \partial H \to \partial G$?*

When such an extension $\hat{i}$ exists, it is called the *Cannon–Thurston map*. In a 1984 preprint, eventually published in 2007 [70], Cannon and Thurston proved that the Cannon–Thurston map exists for the case where G is the fundamental group of a closed fibered hyperbolic 3-manifold and H is the surface subgroup corresponding to the fiber. In 1990s Mitra proved [289, 290] that the Cannon–Thurston map exists in many other situations and posed question (H22) in [291].

In 2013, Baker and Riley [19] constructed an example of a word-hyperbolic group G and a rank 3 free subgroup $F_3 \leq G$ such that the Cannon–Thurston map $\hat{i} : \partial F_3 \to \partial G$ does not exist.

12 Nilpotent groups

*(N1) (A. Myasnikov). *Let G be a free nilpotent group of finite rank. Suppose an element $g \in G$ is fixed by every automorphism of G. Is it true that $g = 1$?*

V. Bludov has communicated the following example of a nontrivial element g of a free nilpotent group of rank 2 and nilpotency class $k \geq 8$, which is fixed by every automorphism, $g = [a, [a, b], [a, b, b], [a, b], \ldots, [a, b]]$, where there are $(2k - 3)$ occurrences of $[a, b]$ after $[a, b, b]$. (Here a and b are generators of the free nilpotent group.)

A. Papistas [322] and, independently, E. Formanek [114] have solved this problem completely by classifying all pairs (r, c) for which $F(r, c)$, the free nilpotent group of rank r and class c, has nontrivial elements fixed by all automorphisms.

*(N2). *Let G be a finitely generated nilpotent group. Is the Dehn function of G equivalent to a polynomial?*

Let $P = \langle X \mid R \rangle$ be a finite presentation of a group G, $F(X)$ a free group on X, and $ncl(R)$ the normal closure of R in $F(X)$. The "area" of $w \in ncl(R)$ is defined by

$$A(w) = \min\left\{ m \mid w = \prod_{i=1}^{m} c_i^{-1} r_i^{e_i} c_i,\ c_i \in F(X), r_i \in R, e_i = \pm 1 \right\}.$$

Now, the *isoperimetric function* of the presentation P is given by

$$\Phi_P(n) = \max\{A(w) \mid w \in ncl(R),\ |w| \leq n\},$$

where $|w|$ is the length of w in $F(X)$.

Let N be the set of all nonnegative integers. For functions $f, h : N \to N$, we define a relation $f \preceq h$ iff there exists a constant K such that $f(n) \leq K \cdot h(Kn) + Kn$ for every $n \in N$. We write $f \simeq h$ iff $f \preceq h$ and $h \preceq f$. It is not hard to show that if P and Q are two finite presentations of a group G, then $\Phi_P \simeq \Phi_Q$. Any function equivalent to Φ_P is called the *Dehn function* of G. From now on, we shall denote the Dehn function of a group G by Φ_G.

S. Gersten [126] proved that for any finitely generated nilpotent group G, Φ_G is bounded by a polynomial of degree 2^h, where h is the Hirsch length of G. G. Conner [83] improved the bound on the degree to 2^c, where c is the nilpotency class of G. C. Hidber [164] proved that $\Phi_G \preceq n^{2c}$. It is known that if G is a free nilpotent group of class c, then $\Phi_G \simeq n^{c+1}$, in particular, Φ_G is equivalent to a polynomial.

In [129], the authors proved that every finitely generated nilpotent group of nilpotency class c admits a polynomial isoperimetric inequality of degree $c + 1$.

However, the problem (N2) was settled in the negative by S. Wenger [378] who proved that there exist finitely generated nilpotent groups G of class 2 whose Dehn function lies between $n^2 d(n)$ and $n^2 \log n$, where $d(n) \to \infty$ as $n \to \infty$.

(N3) (B. I. Plotkin). *Is it true that every locally nilpotent group is a homomorphic image of a torsion-free locally nilpotent group?*

(N4) (G. Baumslag). *Let G be a finitely generated torsion-free nilpotent group. Is it true that there are only finitely many nonisomorphic groups in the sequence Aut(G), Aut(Aut(G)),...?*

We note that J. Hamkins [157] established that every group has a terminating transfinite automorphism tower.

(N5) (G. Baumslag). *Is the property of being directly indecomposable decidable for finitely generated nilpotent groups?*

G. Baumslag, C. F. Miller, and G. Ostheimer [34] answered this problem in the affirmative for *torsion-free* finitely generated nilpotent groups.

(N6) (A. Miasnikov). *Describe all finitely generated nilpotent groups of class 2 that have genus 1. (We say that a group G has genus 1 if every group with the same set of finite homomorphic images as G is isomorphic to G.)*

See the background to the problem (F14).

(N7). *Is every group with an Engel identity $[x, y, \ldots, y] = 1$ locally nilpotent?*

Long time ago, H. Heineken [161] proved that every 3-Engel group (i.e., a group with the identity $[x, y, y, y] = 1$) is locally nilpotent. More recently, G. Havas and M. Vaughan-Lee [160] proved that every 4-Engel group is locally nilpotent.

(N8) (A. Miasnikov).
*(a) *Is it true that equations of the form $[x, y] = g$ ($g \in G$) are decidable in every finitely generated 2-nilpotent group G?*
(b) *Are equations of the form $[x, y] = g$ ($g \in G$) decidable in every finitely generated free nilpotent group?*

V. A. Roman'kov in [339] constructed finitely generated nilpotent groups of nilpotency class 2 where equations of this form are undecidable, thereby giving a negative answer to part (a).

In the same paper, Roman'kov showed that equations of this form are decidable in free nilpotent groups of class 2 of any finite rank. The problem remains open for groups of larger nilpotency class.

(N9) (A. Miasnikov). *Let G be a group. The retract problem in G is the following: given a finitely generated subgroup H of G, decide if H is a retract of G or not.*
(a) *Is the retract problem decidable in every finitely generated nilpotent group G?*
*(b) *Is the retract problem decidable in finitely generated free nilpotent groups?*

V. A. Roman'kov proved in [339] that there is no algorithm that would solve part (a) for any finitely generated nilpotent group of nilpotency class 2. However, the problem of existence of a particular finitely generated nilpotent group with unsolvable retract problem remains open.

In [341], the authors provided a positive answer to part (b).

13 Metabelian groups

Some of the problems about free groups (particularly (F1), (F3)) are also of interest when asked about free metabelian groups. Also, see Problem (A5) in Chapter 8.

(M1). *The isomorphism problem for finitely presented metabelian groups.*

There is an algorithm to determine whether or not a given finitely generated metabelian group is free metabelian; see [148] and [309].

We also note that "most" algorithmic problems about finitely presented metabelian groups are solvable; see [32] and the references therein.

(M2). *Is the automorphism group of a free metabelian group of rank > 3 finitely presented?*

The automorphism group of a free metabelian group of finite rank is known to be finitely generated unless the rank equals 3; see [18] and [17].

(M3) (F. B. Cannonito). *Is there an algorithm that decides whether or not a given finitely presented solvable group is metabelian?*

(M4) (P. Hall). *Are projective groups of infinite countable rank in the class of metabelian groups free metabelian?*

For groups of finite rank, the answer is affirmative; see [10].

(M5) (G. Baumslag). *What can one say about the integral homology of a finitely generated metabelian group?*

For a survey on homological properties of metabelian groups, we refer to [226].

***(M6)** (V. Shpilrain). *Is it true that every IA-automorphism of a free metabelian group of finite rank has a nontrivial fixed point?*

For a background, we refer to [360].
The problem was answered in the negative by M. Kassabov [200].

***(M7)** (R. Goebel). *Is there a group which is not isomorphic to the outer automorphism group of any metabelian group with a trivial center?*

No, there is no such group, see [132].

(M8) (A. Miasnikov). *Let G be the free metabelian group of a finite rank ≥ 2. Is there an algorithm for solving one-variable equations in G?*

Note that the general Diophantine problem in free metabelian groups is undecidable [337].

In [259] and [260], the authors proved that the Diophantine problem for some types of quadratic equations in free metabelian groups is decidable and NP-complete.

14 Solvable groups

(S1) (A. I. Mal'cev). *Describe the automorphism group of a free solvable group of finite rank. In particular, is this group finitely generated?*

The automorphism group of a free solvable group of derived length > 2 and rank > 2 cannot be generated by elementary Nielsen automorphisms, see [152] and [356]. Moreover, every free solvable group of derived length $d > 2$ and rank $r > 2$ has automorphisms that cannot be lifted to automorphisms of the free solvable group of derived length $d + 1$ and the same rank r [358]. It is not known, however, whether or not those automorphism groups are finitely generated. A notable exception is the free metabelian group of rank 3 whose automorphism group is not finitely generated, see [17].

(S2) (M. I. Kargapolov). *The word problem for groups admitting a single defining relation in the variety of all solvable groups of a given derived length.*

We note that the word problem for groups admitting *finitely many* defining relations in the variety of all solvable groups of a given derived length > 2 is, in general, unsolvable; see [203].

(S3) (M. I. Kargapolov). *Is it true that every group of rank > 2 admitting a single defining relation in the variety of all solvable groups of a given derived length has trivial center?*

E. Timoshenko [373] settled this problem in the affirmative for metabelian groups. C. K. Gupta and V. Shpilrain [153] settled the problem (also in the affirmative) for solvable groups of arbitrary derived length, under an additional assumption that the relator is not a proper power modulo any term of the derived series.

(S4) (M. I. Kargapolov). *Is there a number $N = N(k, d)$ such that every element of the commutator subgroup of a free solvable group of rank k and derived length d is a product of N commutators?*

The answer is "yes" for free metabelian groups (see [5]) and for free solvable groups of derived length 3 (see [333]).

(S5) (P. M. Neumann). *Is it true that if A, B are finitely generated solvable Hopfian groups, then $A \times B$ is Hopfian?*

(S6) (V. N. Remeslennikov). *The conjugacy problem for finitely generated abelian-by-polycyclic groups.*

(S7) (G. Baumslag, V. Remeslennikov). *Is a finitely generated free solvable group of derived length 3 embeddable in a finitely presented solvable group?*

*(S8) (B. Fine, V. Shpilrain). *Let u be an element of a group G. We call u a test element if, whenever $\varphi(u) = u$ for some endomorphism φ of the group G, this φ is actually an* *automorphism of G. Does the free solvable group of rank 2 and derived length $d > 2$ have any test elements?*

The most obvious candidate for a test element in a group generated by x and y would be $u = [x, y]$. This, however, is *not* a test element in a free solvable group of derived length $d > 2$, see [154].

V. Roman'kov [338] has constructed test elements in the free solvable group of rank 2 and derived length 3.

It is plausible that the same method can be used for constructing test elements in the free solvable group of any bigger rank as well, but technically it is getting more complicated.

We also mention a related result of E. I. Timoshenko [374] who proved that a free metabelian group of rank > 2 does *not* have any test elements. It was previously known [104] that the free metabelian group of rank 2 does have test elements, for example, $u = [x, y]$.

Finally, a complete solution was given by E. I. Timoshenko in [375]. He showed that the test rank of a free solvable nonabelian group of finite rank is 1 less than the rank of that group. In particular, the free solvable group of rank 2 of any derived length $d > 2$ has test elements.

15 Overview of recent progress in classical areas

In this concluding section, we give a brief survey of recent progress in two "most combinatorial" and classical areas of infinite group theory relevant to our collection, namely free and one-relator groups. Free groups and their properties are at the core of combinatorial group theory; they have been studied for over 100 years now, so it is not surprising that they take the most prominent place in our collection. We try to review, in Section 15.1 below, some of the progress (that is most relevant to our collection) on free groups over the last 20 years or so.

One-relator groups were one of the primary areas of research for Wilhelm Magnus, as well as for Gilbert Baumslag. Both played a crucial role in establishing New York group-theoretic community that we have been privileged to be part of for the last 30+ years. We therefore felt it would be appropriate to give one-relator groups a special treatment in our overview (Section 15.2), especially given that recent progress in this area was rather substantial.

15.1 Recent progress on free groups

Free groups and, in particular, their automorphisms proved to be attractive objects of research, bridging group theory, topology, geometry, and dynamics. Automorphic orbits in free groups were studied by J. H. C. Whitehead in his now classical papers [379] and [380].

Since then, there was a vast amount of research on various properties of automorphic orbits in F_r. We can single out the research avenue focused on fixed points of automorphisms that was popular in the late twentieth century and culminated in a seminal paper by Bestvina and Handel [48].

The early twenty-first century saw a surge in interactions between group theory and theoretical computer science. Most of these interactions are well covered in the monograph [23], but one particular trend that stands out is emerging interest in complexity of group-theoretic algorithms. This includes not only the "traditional" worst-case complexity but also generic- and average-case complexity introduced into group theory in [196] and [194], respectively.

In particular, studies of the complexity of Whitehead's problem (Section 15.1.3) recently led to exciting progress on orbit-blocking words (Section 15.1.1). Then, there are automorphic orbits of potentially positive elements (Section 15.1.5); these, too, saw remarkable progress recently.

In Section 15.1.6, we consider pairs of elements of F_r that have "similar" automorphic orbits; elements like that are called (boundedly) translation equivalent following [193]. Somewhat surprisingly, there was no progress on characterizing or recognizing (boundedly) translation equivalent elements in the last 15 years or so, but we discuss this direction here in the hope that the right tools to attack relevant problems (cf. our Problem (F37)) will eventually be found.

Another direction of research that bridges group theory and theoretical computer science is discrete optimization problems (knapsack problem, subset-sum problem, Post correspondence problem). We discuss these in Section 15.1.7.

We leave out an important direction of studying compressed decision problems in groups because we only include here directions relevant to our collection of open problems. A good source of information on compressed problems in groups is monographs [23, 244] as well as papers [168, 243]. See also the background to our Problem (C1).

15.1.1 Orbit-blocking words

For a given element $w \in F_r$, we say that a word v is w-orbit-blocking (or just orbit-blocking if w is clear from the context) if for any automorphism $\varphi \in \mathrm{Aut}(F_r)$, the word v is not a subword of the cyclic reduction of $\varphi(w)$.

Primitivity-blocking words are the same as x_1-orbit-blocking. The existence of primitivity-blocking words easily follows from Whitehead's observation that the Whitehead graph of any cyclically reduced primitive element is either disconnected or has a cut vertex.

The *Whitehead graph* $\mathrm{Wh}(w)$ of $w \in F_r$ has $2r$ vertices that correspond to the generators and their inverses. For each occurrence of a subword $x_i x_j$ in the word $w \in F_r$, there is an edge in $\mathrm{Wh}(w)$ that connects the vertex x_i to the vertex x_j^{-1}; if w has a subword $x_i x_j^{-1}$, then there is an edge connecting x_i to x_j, etc. There is one more edge (the external edge), namely the edge that connects the vertex corresponding to the last letter of w to the vertex corresponding to the inverse of the first letter.

It was observed by Whitehead himself in his *cut vertex lemma* (see [380]) that the Whitehead graph of any cyclically reduced primitive element w is either disconnected or has a cut vertex, i.e., a vertex that, having been removed from the graph together with all incident edges, increases the number of connected components of the graph. A short and elementary proof of this result was recently given in [163], and a more general case of primitive elements in subgroups of F_r was recently treated in [12].

Thus, for example, if the Whitehead graph of w (without the external edge) is complete (i.e., any two vertices are connected by at least one edge), then w is primitivity-blocking because in this case, if w is a subword of u, then the Whitehead graph of u, too, is complete and therefore is connected and does not have a cut vertex. Here are some examples of primitivity-blocking words: $x_1^n x_2^n \ldots x_r^n x_1$ (for any $n \geq 2$), $[x_1, x_2][x_3, x_4] \ldots [x_{n-1}, x_n] x_1^{-1}$ (for an even n), etc. Here $[x, y]$ denotes $x^{-1} y^{-1} xy$.

In [177], it was shown that for any $w \in F_2$, there are w-orbit-blocking words. This was based on an explicit description of bases (equivalently, of automorphisms) of F_2 from [80]. Then, in [218], the following general result was established:

Theorem 1. *Let $w \in F_r$ have cyclically reduced length ℓ. Let $\{v_i\}_{i=1}^{\ell+1}$ be a sequence of primitivity-blocking words such that there is no cancellation between adjacent terms of the sequence, nor between $v_{\ell+1}$ and v_1. Then $\prod_{i=1}^{\ell+1} v_i$ is w-orbit-blocking.*

Note that, interestingly, w-orbit-blocking words described in Theorem 1 depend on the length of w but not on the actual w.

As can be seen from the statement of Theorem 1, primitivity-blocking words play an important role, so we are going to discuss them separately.

15.1.2 Primitivity-blocking words

By definition, primitivity-blocking words are those that cannot be subwords of any *cyclically reduced* primitive elements. What is needed to prove Theorem 1 is words with somewhat stronger property: they cannot be subwords even of some primitive elements that are not cyclically reduced.

Definition 15.1. Call a basis of F_r *strongly reduced* if there is no element $g \in F_r$ such that conjugating each element of the basis by g decreases the sum of the lengths of the basis elements.

Then there is the following technical result of independent interest.

Lemma 15.2 ([218]). *Let B be a strongly reduced basis of F_r. No primitivity-blocking word appears as a subword of any element in B.*

Then, to show that blocking primitivity does not have to be due to connectivity properties of the Whitehead graph, we include the following examples.

Proposition 15.3 ([218]). **(a)** *In the group F_2, the words $x_1^k x_2^k$ are primitivity-blocking for any $k \geq 2$.*
(b) *In the group F_r, $r \geq 3$, the words $x_1^k \ldots x_r^k$ are not primitivity-blocking for any $k \geq 1$.*

Part (b) of Proposition 15.3 can be generalized as follows:

Proposition 15.4 ([218]). *Let F_r where $r \geq 3$ be generated by the set $X \sqcup Y$. Let $\langle X \rangle$ and $\langle Y \rangle$ denote the subgroups of F_r generated by X and Y, respectively. Let $w = w_X w_Y$, where $w_X \in \langle X \rangle$, $w_Y \in \langle Y \rangle$. Then w is not a primitivity-blocking word.*

We note that the shortest primitivity-blocking word in F_2 is $x_1^{-1} x_2 x_1$. The Whitehead graph (with or without the external edge) of this word does have a cut vertex though.

It is natural then to look for the shortest primitivity-blocking words in F_r for $r > 2$.

Theorem 15.5 ([218]). *The word $w = x_1 x_2 x_3 \ldots x_{r-1} x_r^2 x_{r-1} \ldots x_3 x_2 x_1^{-1} \in F_r$ is the shortest primitivity-blocking word for $r > 2$.*

A similar word $w = x_1 x_2 \ldots x_{r-1} x_r^2 x_{r-1} \ldots x_2 x_1$ is not primitivity-blocking in any F_r, $r \geq 2$. Indeed, for $r = 2$, w is a subword of $x_2 x_1 x_2^2 x_1$, which is a cyclically reduced primitive word.

For $r > 2$, let $u = x_2 x_3 \ldots x_{r-1} x_r^2 x_{r-1} \ldots x_3 x_2$. Consider the word $v = u x_1 u x_1 x_3$. This v clearly contains w as a subword and is cyclically reduced. Apply the automorphism $\varphi = x_1 \mapsto u^{-1} x_1$, $x_i \mapsto x_i$, $i \geq 2$. Then $\varphi(v) = x_1^2 x_3$ is a primitive element, and therefore so is v.

Finally, we mention that primitivity-blocking words in F_2 can be algorithmically recognized [218]. For F_r, $r > 2$, the problem is open.

15.1.3 Whitehead's automorphism problem

Whitehead's automorphism problem is as follows: given $u, v \in F_r$, decide whether or not v is an automorphic image of u. The problem is solved by *Whitehead's algorithm* that dates back to 1936 [379]:

(1) First reduce u to a word of minimal possible length by applying "elementary" Whitehead automorphisms. The same for v. This part is "greedy" and has quadratic time complexity with respect to the length of the input.
(2) If the reduced words are not of the same length, then u and v are not in the same automorphic orbit. If they are of the same length, then things get interesting (in terms of complexity).

A challenging open problem is to determine the worst-case time complexity of Whitehead's problem in F_r.

In F_2, the worst-case complexity was shown to be at most quadratic [302]. In fact, if the length of $u \in F_2$ is m and it cannot be reduced by any Whitehead automorphism, then the maximum possible number of elements of length m in the automorphic orbit of u is precisely $8m - 40$ for $m \geq 10$, see [201].

In F_3, the maximum number is $48m^3 - 480m^2 + 1104m - 672$ for $m \geq 11$ as suggested (but not proved) by C. Sims based on computer experiments. A particular Whitehead-reduced word of length m whose automorphic orbit (limited to elements of length m) has the cardinality given by the latter polynomial is $u = x_1^k x_2 x_1 x_2^{-1} x_1 x_2^2 x_3^2$, where $k = m - 8$.

D. Lee [230] came close to solving this problem completely. Specifically, she proved that the cardinality of $A(u)$ is bounded by a polynomial function of $|u|$ under the following condition: If two letters x_i (or x_i^{-1}) and x_j (or x_j^{-1}) with $i < j$ occur in u, then the total number of $x_i^{\pm 1}$ occurring in u is strictly less than the total number of $x_j^{\pm 1}$ occurring in u.

Then, D. Lee [231] proved that, under the same assumption on u, the cardinality of $A(u)$ is bounded by a polynomial function of $m = |u|$ of degree $2r - 3$, and that this bound is sharp.

15.1.4 Generic- and average-case complexity of Whitehead's problem

Informally, the generic-case complexity of an algorithm is complexity on "most" inputs, or on "random" inputs. For a formal treatment, see [196].

The generic-case complexity of the classical Whitehead's algorithm (see above) is linear, as shown in [197]. The reason is that for a word u selected uniformly at random from the set of elements of length n in F_r, applying any Whitehead's automorphism (except permutations on the set of generators and their inverses) increases the length of u with probability $1 - O(2^{-n})$.

To get a meaningful (e. g., subexponential) estimate of the average-case complexity of Whitehead's problem, one has to get a more precise (than just "exponential") upper bound on the worst-case complexity. It may be then possible to get a subexponential bound on the average-case complexity of Whitehead's problem without getting a subexponential bound on the worst-case complexity.

In [177, 218, 362], the following variant of Whitehead's problem was considered: given a fixed $u \in F_r$, decide, on an input $v \in F_r$ of length n, whether or not v is an automorphic image of u. Thus, in this variant the input consists of just one word.

It turns out that the average-case complexity of this version of the Whitehead problem is constant if the input v is a cyclically reduced word. For a formal definition of the average-case complexity of an algorithm in the context of group theory, we refer the reader to [194].

The algorithm with a constant average-case complexity from [218] is a combination of two different algorithms running in parallel: one is fast but may be inconclusive, whereas the other is conclusive but relatively slow. This idea has been used for group-theoretic algorithms since at least [194].

Before running the two algorithms, a precomputation would be performed on u. Specifically, u is reduced to an element of minimum length in its automorphic orbit. This takes worst-case quadratic time in $|u|$. However, since u is a fixed word, this amounts to a constant time for the algorithm. Denote the obtained element of minimum length in the orbit of u by $\bar{u}$. Once the word's length has been minimized, the two algorithms will be run in parallel on the result.

A fast algorithm $\mathcal{T}$ would detect a $\bar{u}$-orbit-blocking subword $B(\bar{u})$ of a (cyclically reduced) input word v, as follows. Let n be the length of v. The algorithm $\mathcal{T}$ would read the initial segments of v of length k, $k = 1, 2, \ldots$, adding one letter at a time, and check if this initial segment has $B(\bar{u})$ as a subword. This takes time bounded by $C \cdot k$ for some constant C; see, e. g., [216, p. 338].

The "usual" Whitehead algorithm, call it $\mathcal{W}$, would minimize $|v|$ taking time quadratic in $|v|$. Denote the obtained element of minimum length in the orbit of v by $\bar{v}$. If $|\bar{v}| \neq |\bar{u}|$, then $\mathcal{W}$ stops and reports that v is not in the automorphic orbit of u. If $|\bar{v}| = |\bar{u}|$, then the algorithm $\mathcal{W}$ would apply all possible sequences of elementary Whitehead automorphisms that do not change the length of $\bar{v}$ to see if any of the resulting elements are equal to $\bar{u}$. This part may take exponential time in $|\bar{v}| = |\bar{u}|$, but since we consider $|u|$ constant with respect to $|v|$ and $|u|$ bounds $|\bar{u}|$ above, exponential time in $|\bar{u}|$ is still constant with respect to $|v|$. Thus, the total time that the algorithm $\mathcal{W}$ takes is quadratic in $|v|$.

Finally, the algorithm $\mathcal{A}$ will consist of the algorithms $\mathcal{T}$ and $\mathcal{W}$ running in parallel. Then we have:

Theorem 15.6 ([218]). *Suppose possible inputs of the above algorithm $\mathcal{A}$ are cyclically reduced words that are selected uniformly at random from the set of cyclically reduced words of length n. Then the average-case time complexity (i. e., expected runtime) of the algorithm $\mathcal{A}$, working on a classical Turing machine, is $O(1)$, a constant that does not depend on n.*

15.1.5 Potentially positive elements

An element u of F_r is called *positive* if no x_i occurs in u to a negative exponent. An element u is called *potentially positive* if $\alpha(u)$ is positive for some automorphism α of the group F_r.

An element u is called *stably potentially positive* if $\alpha(u)$ is positive for some automorphism α of the group F_n for some $n \geq r$.

The motivation for considering potentially positive elements comes from the fact that various properties of one-relator groups are easier to establish if the relator is a positive word. For example, Baumslag [28] showed that one-relator groups with a positive relator are residually solvable, and Wise [384] showed that one-relator groups with a positive relator (satisfying a small cancellation condition) are residually finite. All these properties obviously hold upon replacing "positive" with "potentially positive" or even "stably potentially positive."

Clark and Goldstein [75] proved that all stably potentially positive elements are potentially positive.

Goldstein [133] and D. Lee [234] offered (exponential time) algorithms for deciding potential positivity in F_2. Another algorithm was offered by Silva and Weil [363].

Recently, Koch-Hyde, O'Connor, and Olive [217] reported an algorithm with the worst-case time complexity $O(n^2)$. This algorithm has linear generic-case complexity. If inputs are cyclically reduced, then the average-case complexity of this algorithm is constant. This is because, as shown in [217], there are potential positivity-blocking words in F_2, i. e., words that cannot be subwords of any potentially positive word. An example would be $xyx^{-1}y^{-1}x$.

A very interesting and challenging question is (see our Problem (F33)(c)): how many potentially positive elements of length n are there in F_r?

The set of positive elements is exponentially negligible in the set of potentially positive elements because any element with only positive occurrences of all but one of the generators is potentially positive. (Note that the number of positive elements of length n in F_r is r^n, whereas the total number of elements of length n is $2r \cdot (2r-1)^n$.)

The number of potentially positive elements of length n is bounded from below by the number of elements of length n with only positive occurrences of all but one of the generators. The latter number is $> (\frac{r^2+2r-1}{r+1})^n$.

For $r > 3$, a larger lower bound is provided by the number of primitive elements of length n, which is $O(n(2r-3)^n)$ by [330].

Recently, [101] obtained a tight estimate for the growth rate in the case $r = 2$ by proving that the number of potentially positive elements of length n in F_2 is $O((\lambda + \epsilon)^n)$ for any $\epsilon > 0$, where $\lambda \approx 2.505$ is the largest root of the polynomial $\lambda^4 - 3\lambda^3 + \lambda^2 + \lambda - 1$.

15.1.6 Translation equivalence

Two elements u, v of a free group F_r are called *translation equivalent* if for every free and discrete isometric action of F_r on an $\mathbb{R}$-tree, translation lengths of u and v are equal.

A purely combinatorial characterization of translation equivalence was given in [193]: elements u, v are translation equivalent if the cyclic length of $\varphi(u)$ equals the cyclic length of $\varphi(v)$ for every automorphism φ of the group F_r.

A ramification of this definition is: two elements u, v are called *boundedly translation equivalent* if the ratio of the cyclic lengths of $\varphi(u)$ and $\varphi(v)$ is bounded away from 0 and from ∞ when φ runs through all automorphisms of F_r.

It is natural to ask (see our problem) if there is an algorithm which, when given two elements of a finitely generated free group, decides whether or not they are (boundedly) translation equivalent.

It is not quite trivial to even produce examples of (boundedly) translation equivalent elements. In [193], it was shown that in F_2, any $u(x, y)$ is translation equivalent to $u(x, y)$ read backwards.

Another source of translation equivalent elements comes from *character equivalence*. We say that u and v are *trace equivalent*, or *character equivalent*, if for every representation $\alpha : F_r \to SL_2(\mathbb{C})$ one has $tr(\alpha(u)) = tr(\alpha(v))$. Character equivalent words come from the so-called "trace identities" in $SL_2(\mathbb{C})$ and are quite plentiful (see, for example, [169]).

D. Lee showed [232] that whenever g is translation equivalent to h in F_r and $w(x, y)$ is arbitrary, one has $w(g, h)$ translation equivalent to $w(h, g)$ in F_r. She also reported an algorithm that decides translation equivalence in F_2.

In [233], D. Lee also offered an algorithm that decides bounded translation equivalence in F_2.

Surprisingly, there has not been any progress on (bounded) translation equivalence since then.

15.1.7 Discrete optimization problems

Many classical discrete optimization problems can be generalized and studied in noncommutative groups. For example, problems concerning integers (subset-sum, knapsack problem, etc.) make perfect sense when the group of additive integers is replaced by an arbitrary (noncommutative) group G. The classical lattice problems are about subgroups (integer lattices) of the additive groups $\mathbb{Z}^n$ or $\mathbb{Q}^n$, whereas their noncommutative versions deal with arbitrary finitely generated subgroups of a group G. The traveling salesman problem and the Steiner tree problem make sense for arbitrary finite subsets of vertices in a given Cayley graph of a noncommutative infinite group (with the natural graph metric). The Post correspondence problem carries over in a straightforward fashion from a free monoid to an arbitrary group (cf. our Problem (F41)). This list of examples can be easily extended, but the point here is that many classical discrete optimization problems have natural and interesting noncommutative versions.

Suppose that a group G is specified by a countable generating set X and that elements of G are specified as group words over X. Given $g_1, \ldots, g_k, g \in G$, the *subset-sum problem* asks whether $g = g_1^{\epsilon_1} \cdots g_k^{\epsilon_k}$ for some $\epsilon_1, \ldots, \epsilon_k \in \{0, 1\}$.

The *knapsack problem* asks whether the equality $g = g_1^{\epsilon_1} \cdots g_k^{\epsilon_k}$ holds for some nonnegative integers $\epsilon_1, \ldots, \epsilon_k \in \{0, 1\}$.

For the subset-sum problem, [299] shows that there is a polynomial-time algorithm solving the problem for hyperbolic groups (including free groups). The algorithm constructs finite automata over hyperbolic groups via two operations, completion and folding. The authors prove, however, that the problem is NP-complete for the following two families of groups: (i) free metabelian groups of finite rank ≥ 2; and (ii) wreath products of finitely generated infinite abelian groups. It remains NP-complete for Thompson's group F and Baumslag's group [29].

For the knapsack problem, [299] shows that there is also a polynomial-time algorithm solving the problem for hyperbolic groups. See also [245] for refinements of this result.

The classical *Post correspondence problem* (PCP) can be formulated in algebraic terms as follows: given free monoids M, N of finite rank and two homomorphisms φ, ψ from M to N, decide whether there exists a nonidentity element $w \in M$ such that $\varphi(w) = \psi(w)$ in N. In general, PCP is undecidable [326].

If $M = F_n$ and $N = F_m$ are free groups, then PCP is still open (cf. our Problem (F41)). In the context of groups, this problem was first introduced and studied in [298]. See also [74] for variations of this problem for free groups, and [73] for adaptations in other groups including hyperbolic and virtually nilpotent groups.

15.2 Recent progress on one-relator groups

Major progress in the study of one-relator groups occurred in the last 20 years or so. In this section we will give an overview of some of the main recent developments in the subjects, concentrating on those that more directly address problems raised by Baumslag and related to his work. For a more comprehensive and in depth discussion of important recent developments in the study of one-relator groups, we refer the reader to a survey by Linton and Nyberg-Brodda [240].

Recent progress in the theory of one-relator groups was made possible by the development of several new tools and ideas in geometric group theory and low-dimensional topology.

15.2.1 Virtually special groups and quasiconvex hierarchies

The first tool is the theory of *virtually special* groups and *quasiconvex hierarchies*, originally developed by Wise [387], and elaborated and clarified by Agol, Groves, and Manning [3], ultimately leading to the proof by Agol [1] of the long standing virtually Haken conjecture in 3-manifold topology. A finitely generated group G is *virtually special* of G has a subgroup of finite index H which admits a properly discontinuous cocompact isometric action on a *special* CAT(0) cubical complex. A key result of the theory shows that in such a situation H embeds in a right-angled Artin group (RAAG) [156]. Hence virtually special groups are linear (over $\mathbb{R}$) and therefore residually finite. Moreover, if G is word-hyperbolic and virtually special then any quasiconvex subgroup of G is separable in G [156].

The "quasiconvex hierarchies" machinery provides a rich source of constructing virtually special word-hyperbolic (and relatively hyperbolic) groups. For a word-hyperbolic group G, a *quasiconvex hierarchy* $\mathcal{H}$ is an iterated system of finite graph-of-groups decompositions or splittings, starting from G and iteratively splitting the vertex groups, where all the vertex and edge groups are word-hyperbolic, the edge groups are quasiconvex in G and some additional technical conditions, depending on the version of the notion, are required to be satisfied. The hierarchy terminates with subgroups that are either trivial, finite, free, virtually free, virtually special, etc., again depending on the version of the notion. Wise's quasiconvex hierarchy theorem states that if a word-hyperbolic group G admits a quasiconvex hierarchy terminating in trivial groups then G is virtually special. We refer the reader to [1, 2, 43, 224, 387] for more details and additional references on the virtually Haken conjecture, quasiconvex hierarchies, virtually special groups, special cube complexes, and related topics.

It turns out that the quasiconvex hierarchy machinery applies to many one-relator groups (both hyperbolic and relatively hyperbolic) by adapting the classic Magnus–Moldavanskii one-relator hierarchy. The Magnus–Moldavanskii hierarchy, starting with a finitely generated one-relator group $G = \langle X \mid w = 1 \rangle$, embeds $G = G_0$ as a subgroup $G \leq G'$ of another one-relator group G', such that G' splits as an HNN-extension $G' = G_1 *_\psi$ where G_1 is a one-relator group with the defining relation shorter than w, and where $\psi : A \to B$ is an isomorphism between two (free) Magnus subgroups of G_1. This process is then iteratively applied to G_1 and so on, until one reaches the trivial group or the finite cyclic group (depending on whether or not w is a proper power in $F(X)$). There is a useful simplification of this process, due to an unpublished argument of Masters (see the recent paper of Linton [239] for details) which shows that one can dispense with embedding $G = G_0$ into G' and directly decompose $G = G_0$ as $G_0 = G_1 *_\psi$ as above, and iterate.

In [387], Wise showed that for one-relator groups with torsion the Magnus–Moldavanskii hierarchy is quasiconvex and used the quasiconvex hierarchy theorem to fully resolve (OR1)(a):

Theorem 15.7. *Let $G = \langle x_1, \ldots, x_m \mid w^n = 1 \rangle$ where $m \geq 1$, $n \geq 2$, and w is a nontrivial cyclically reduced word. Then G is word-hyperbolic, virtually special, residually finite, and linear over $\mathbb{R}$.*

Linton later used the Magnus–Moldavanskii hierarchy for arbitrary word-hyperbolic one-relator groups to prove [238]:

Theorem 15.8. *Let $G = \langle x_1, \ldots, x_m \mid w = 1 \rangle$ where $m \geq 1$ and w is a nontrivial cyclically reduced word, and suppose that G is word-hyperbolic. Let $Y \subset \{x_1, \ldots, x_m\}$ be a subset omitting some generator x_i such that x_i or x_i^{-1} appears in w and let $A = \langle Y \rangle \leq G$ be the Magnus subgroup of G generated by Y. Then A is quasiconvex in G.*

Another application, due to Linton, which also uses the negative immersions technique, is discussed in Theorem 15.18 below.

By a result of Martinez-Pedroza and Wise [274], if $G = \langle X \mid r^m = 1 \rangle$, where X is finite, r is a nontrivial cyclically reduced word in X, and $m \geq 3|r|$, then G is a locally quasiconvex word-hyperbolic group. Since, by Theorem 15.7, G is virtually special, and virtually special word-hyperbolic groups are separable with respect to quasiconvex subgroups, this yields:

Corollary 15.9. *Let $G = \langle X \mid r^m = 1 \rangle$, where X is finite, r is a nontrivial cyclically reduced word in X, and $m \geq 3|r|$. Then G is word-hyperbolic and subgroup separable.*

Corollary 15.9 is essentially due to Wise [387] although does not appear to be explicitly stated there.

A powerful general result of Agol, obtained by applying the quasiconvex hierarchies machinery, says that if G is a word-hyperbolic group admitting a properly discontinuous cocompact isometric action on CAT(0) cubical complex then G is virtually special. This result applies to all $C'(1/6)$ and $C'(1/4) - T(4)$ small cancellation finitely presented groups, which are known to admit above type actions on CAT(0) cubical complexes by a result of Wise [385], and in particular to one-relator groups of this kind, yielding:

Theorem 15.10. *Let $G = \langle x_1, \ldots, x_m \mid w = 1 \rangle$, where $m \geq 1$, w is a nontrivial cyclically reduced word, and this presentation, after symmetrization, satisfies either $C'(1/6)$ or $C'(1/4) - T(4)$ small cancellation condition.*

Then G is word-hyperbolic, virtually special, residually finite, and linear over $\mathbb{R}$.

Theorem 15.10 is essentially due to Agol [1], although also not explicitly stated there.

15.2.2 Nonpositive immersions and coherence

Another key technique, more specifically applicable to one-relator presentations, comes from the notion of "a complex with nonpositive immersions" and its various generalizations. This notion was first introduced by Wise specifically to study problem (OR1)(b) about coherence of one-relator groups. A finite 2-complex X has *nonpositive immersions* if for every compact connected complex Y and a combinatorial immersion of Y into X, either $\chi(Y) \leq 0$ or Y is contractible. Similarly, a finite 2-complex X has *negative immersions* if for every compact connected complex Y and a combinatorial immersion of Y into X, either $\chi(Y) < 0$ or Y is contractible. Louder and Wilton later gave slight relaxations of these notions, yielding the same results, where the "or Y is contractible" part is replaced by the condition that Y be reducible to a graph by a sequence of moves through free faces. The negative immersion property for X also has a uniform version where the $\chi(Y) < 0$ condition is made uniform in a certain normalized sense.

When introducing the notion of nonpositive immersions, Wise showed that it can be thought of as an analog of both topological and curvature features of hyperbolic 3-manifolds and produced a proof that fundamental groups of 2-complexes with nonpositive immersions are coherent. Wise then provided an argument that for a one-relator group $G = \langle x_1, \ldots, x_m \mid w = 1 \rangle$, where w is a positive word, the presentation complex has nonpositive immersions if w is not a proper power or has a finite cover with nonpositive immersions if w is a proper power, and hence G is coherent for any positive w. This result naturally led Wise to conjecture that presentation complexes of one-relator groups have negative immersions (up to finite covers, in the case of one-relator groups with torsion). However, it later turned out that Wise's original proof that fundamental groups of 2-complexes with nonpositive immersions are coherent contains a gap and this gap remains unfilled. Thus the statement below remains a conjecture:

Conjecture 15.11 (Wise). Let X be a finite 2-complex with nonpositive immersions. Then $\pi_1(X)$ is coherent.

It was later shown by Helfer and Wise [162] and, independently, by Louder and Wilton [249] that the presentation complex of any torsion-free one-relator group has nonpositive immersions.

While Conjecture 15.11 remains open, the notions of nonpositive immersions and its variants, as well as negative immersions and uniformly negative immersions, turned out to be quite useful in the study of coherence of one-relator groups and of related questions. Thus Louder and Wilton [250] and, independently, Wise [388] were able to use these ideas to prove that one-relator groups with torsion are coherent:

Theorem 15.12. Let $G = \langle x_1, \ldots, x_m \mid w^n = 1 \rangle$, where $m \geq 1$, $n \geq 2$, and w is a nontrivial cyclically reduced word. Then G is coherent.

Both proofs exploited stronger versions of nonpositive immersions for finite covers of the presentation complex of G.

Subsequently Jaikin-Zapirain and Linton exploited a different, more algebraic, approach to the above conjectures. Using a combination of deep homological, group algebra, and analytic techniques they were able to establish a weaker version of Conjecture 15.11 above [188]:

Theorem 15.13. *Let X be a finite 2-complex with nonpositive immersions. Then $\pi_1(X)$ is homologically coherent.*

Recall that a group G is *homologically coherent* if every finitely generated subgroup of G is of type $FP_2(\mathbb{Z})$. The property of being $FP_2(\mathbb{Z})$ is strictly weaker than being finitely presented. Jaikin-Zapirain and Linton [188] were able to promote their method to completely solve (OR1)(a):

Theorem 15.14. *Let $G = \langle x_1, \ldots, x_m \mid w = 1 \rangle$, where $m \geq 1$ and w is a nontrivial cyclically reduced word, and let K be a field of characteristic 0. Then the group G is coherent and the group algebra $K[G]$ is coherent.*

The properties of a 2-complex X having negative immersions and uniformly negative immersions turned out to be particularly closely related to the subgroup structure of one-relator groups, and they also turned out to be related to the Gersten conjecture, (O6), and to Problem (OR7). Louder and Wilton showed that for presentation complexes of one-relator groups, the conditions of having negative immersions and uniformly negative immersions are equivalent, so there is no need to distinguish between the two notions.

Louder and Wilton proved the following [251]:

Theorem 15.15. *Let $G = \langle x_1, \ldots, x_m \mid w = 1 \rangle$, where $m \geq 1$ and w is a nontrivial cyclically reduced word, and let X be the presentation complex of G. Then the following conditions are equivalent:*
1. *Complex X has negative immersions.*
2. *Complex X has uniformly negative immersions.*
3. *Every two-generator subgroup of G is free.*
4. *The primitivity rank $\pi(w) > 2$.*

For a nontrivial element w of the free group F_m, the *primitivity rank* $\pi(w)$ is the smallest rank of a subgroup $H \leq F_m$ containing w as a nonprimitive element; and $\pi(w) = \infty$ if no such H exists. The notion of primitivity rank was originally introduced by Puder [328] in the study of word measures on finite groups, but it also proved highly useful in the study of one-relator groups.

Louder and Wilton later also proved [252]:

Theorem 15.16. *Let $G = \langle x_1, \ldots, x_m \mid w = 1 \rangle$, where $m \geq 1$ and w is a nontrivial cyclically reduced word, and let X be the presentation complex of G. Suppose that G has negative immersions. Then:*
1. *Every finitely generated one-ended subgroup of G is co-Hopf.*

2. *For every $r \geq 1$, G has only finitely many conjugacy classes of subgroups H such that H is finitely generated, one-ended and has abelianization of torsion-free rank at most r.*
3. *Every finitely generated noncyclic subgroup of G is large, that is, has a subgroup of finite index admitting an epimorphism on F_2.*

Since, by the above result, every two-generator subgroup in a one-relator group $G = \langle x_1, \ldots, x_m \mid w = 1\rangle$ with negative immersions is free, G cannot contain Baumslag–Solitar subgroups. In view of Gersten's conjecture (O6), this implication led Louder and Linton to pose the following [251, Conjecture 1.9]:

Conjecture 15.17. Every one-relator group with negative immersions is word-hyperbolic.

In [239], Linton used a combination of the negative immersions methods and the techniques of quasiconvex hierarchies and virtually special groups and complexes to establish the following strengthened version of this conjecture:

Theorem 15.18. *Let $G = \langle x_1, \ldots, x_m \mid w = 1\rangle$, where $m \geq 1$ and w is a nontrivial cyclically reduced word, and let X be the presentation complex of G. Suppose that G has negative immersions.*

Then G is word-hyperbolic and virtually special.

In [239], Linton derived several useful corollaries of Theorem 15.18 for one-relator groups with negative immersions. Thus he proved that every such group G is residually finite and linear over $\mathbb{R}$, and that all finitely generated subgroups of G are word-hyperbolic. (The last conclusion follows from Theorem 15.18 since G has virtual cohomological dimension 2, is coherent, and, by a result of Gersten, all finitely presented subgroups of word-hyperbolic groups of virtual cohomological dimension 2 are word-hyperbolic.)

Theorem 15.18 of Linton remains the strongest result in the direction of Gersten's conjecture (Problem (O6)) known to date.

16 Hall of fame

These are names (in the alphabetical order) of people who have published solutions of one or more problems from our list after the first draft was put online in June 1997:
- Y. Antolín, A. Jaikin-Zapirain – Problem (F29)
- O. Baker, T. R. Riley – Problem (H22)
- F. Bassino, C. Nicaud, P. Weil – Problem (C6)(b)
- V. G. Bardakov, R. Mikhailov – Problem (F32)(b)
- L. Bartholdi – Problem (G3)
- J. Behrstock, C. Drutu; J. Tao – Problem (C3)
- M. C. Bell, S. Schleimer – Problem (B13)
- M. Bestvina, N. Brady – Problem (FP3)
- M. Bestvina, M. Feighn, M. Handel – Problem (F2)
- S. Bigelow – Problems (O10), (B1)
- V. Bludov – Problem (N1)
- O. Bogopolski, O. Maslakova – Problem (F1)(a)
- A. V. Borshev, D. I. Moldavanskii – Problem (OR9)
- M. Bridson, D. Wise – Problem (H14)
- M. Burger, S. Mozes – Problem (F21)
- L. Chen, Y. Lodha – Problem (FP11)
- A. Clark, R. Goldstein – Problem (F33)(b)
- A. Clifford, R. Goldstein – Problem (F38)(b)
- T. Coulbois, A. Khelif – Problem (E3)
- F. Dahmani, V. Guirardel – Problem (H3)
- A. Darbinyan – Problem (A2)
- M. Evans – Problem (MA2)
- M. Feighn, M. Handel – Problem (FP8)
- E. Formanek – Problem (N1)
- J. Friedman – Problem (O9)
- G. Gardam – Problem (O12)(b)
- R. Goebel, A. Paras – Problem (M7)
- J. González-Meneses – Problem (B11)
- J. Howie – Problem (FP12)
- S. Humphries – Problem (B7)(a), (b)
- S. V. Ivanov – Problem (FP4)
- A. Jaikin-Zapirain – Problems (F12)
- A. Jaikin-Zapirain, M. Linton – Problem (OR1)(c)
- M. Kapovich – Problem (H2)
- M. Kassabov – Problem (M6)
- B. Khan – Problem (F25)(b)
- O. Kharlampovich, A. Myasnikov – Problems (O8)(a), (b), (F10), (F21), (OR16), and (FP7)

- L. Koch-Hyde, S. O'Connor, E. Olive, V. Shpilrain – Problem (F39)
- K. Kordek, D. Margalit – Problems (B6)(b), (B8)
- D. Krammer – Problems (O10), (B1)
- D. Lee – Problems (F3), (F7), and (F37)(b)
- G. Levitt, A. Myasnikov, V. Shpilrain – Problem (F4)
- P. Linnell, T. Schick – Problem (B7)(a), (b)
- J. McCool – Problem (F16)
- I. Mineyev – Problem (O9)
- V. Nekrashevych – Problem (G4)
- A. Olshanskii – Problems (OR7)(b), (c)
- S. Orevkov – Problem (B8)(a)
- D. Osin – Problem (FP18)
- A. Papistas – Problem (N1)
- D. Puder, C. Wu – Problem (F17)
- C. F. Rocca, Jr., E. Turner – Problem (FP13)
- V. Roman'kov – Problems (N8)(a), (S8)
- V. A. Roman'kov, N. G. Khisamiev, A. A. Konyrkhanova – Problem (N9)(b)
- S. Schleimer – Problem (C1)
- Z. Sela – Problem (O8)(a)
- R. Sharp – Problem (G8)
- I. Snopce, S. Tanushevski, P. Zalesskii – Problem (F11)
- E. I. Timoshenko – Problems (FP13), (S8)
- S. Wenger – Problem (N2)
- J. S. Wilson – Problem (G3)
- D. Wise – Problems (OR1), (FP16)

Bibliography

[1] I. Agol, *The virtual Haken conjecture*, Doc. Math. **18** (2013), 1045–1087. With an appendix by I. Agol, D. Groves, and J. Manning.

[2] I. Agol, *Virtual properties of 3-manifolds*, in: *Proceedings of the International Congress of Mathematicians—Seoul 2014, Vol. 1*, 141–170, Kyung Moon Sa, Seoul, 2014.

[3] I. Agol, D. Groves and J. F. Manning, *An alternate proof of Wise's malnormal special quotient theorem*, Forum Math. Pi **4** (2016), e1, 54 pp.

[4] S. Akbulut and R. Kirby, *A potential smooth counterexample in dimension 4 to the Poincare conjecture, the Schoenflies conjecture, and the Andrews–Curtis conjecture*, Topology **24** (1985), 375–390.

[5] Kh.S. Allambergenov and V. A. Romankov, *Products of commutators in groups*, Dokl. Akad. Nauk UzSSR **4** (1984), 14–15 (Russian).

[6] J. M. Alonso, *Combings of groups*, in: *Algorithms and Classification in Combinatorial Group Theory* (Berkeley, CA, 1989), Math. Sci. Res. Inst. Publ., **23**, 165–178, Springer, New York, 1992.

[7] J. M. Alonso, T. Brady, D. Cooper, V. Ferlini, M. Lustig, M. Mihalik, M. Shapiro and H. Short, *Notes on word hyperbolic groups*, in: *Group Theory from a Geometrical Viewpoint* (Trieste, 1990), 3–63, World Sci. Publ., River Edge, NJ, 1991, ISBN 981-02-0442-6.

[8] J. J. Andrews and M. L. Curtis, *Free groups and handlebodies*, Proc. Am. Math. Soc. **16** (1965), 192–195.

[9] Y. Antolín and A. Jaikin-Zapirain, *The Hanna Neumann conjecture for surface groups*, Compos. Math. **158** (2022), 1850–1877.

[10] V. A. Artamonov, *Projective metabelian groups and Lie algebras*, Izv. Akad. Nauk SSSR, Ser. Mat. **42** (1978), 226–236, 469 (Russian).

[11] G. Arzhantseva and A. Ol'shanskii, *Genericity of the class of groups in which subgroups with a lesser number of generators are free*, Mat. Zametki **59**(4) (1996), 489–496 (Russian).

[12] D. Ascari, *A fine property of Whitehead's algorithm*, Groups Geom. Dyn. **18** (2024), 235–264.

[13] C. J. Ashcroft, *Property (T) in density-type models of random groups*, Groups Geom. Dyn. **17** (2023), 1325–1356.

[14] M. Auslander and R. C. Lyndon, *Commutator subgroups of free groups*, Am. J. Math. **77** (1955), 929–931.

[15] S. Bachmuth, *Braid groups are linear groups*, Adv. Math. **121** (1996), 50–61.

[16] S. Bachmuth and H. Mochizuki, $E_2 \neq SL_2$ *for most Laurent polynomial rings*, Am. J. Math. **104** (1982), 1181–1189.

[17] S. Bachmuth and H. Mochizuki, *The nonfinite generation of Aut(G), G free metabelian of rank* 3, Trans. Am. Math. Soc. **270** (1982), 693–700.

[18] S. Bachmuth and H. Mochizuki, $Aut(F) \to Aut(F/F'')$ *is surjective for free group F of rank* ≥ 4, Trans. Am. Math. Soc. **292** (1985), 81–101.

[19] O. Baker and T. R. Riley, *Cannon–Thurston maps do not always exist*, Forum Math. Sigma **1** (2013), e3.

[20] V. G. Bardakov and R. Mikhailov, *On certain questions of the free group automorphisms theory*, Commun. Algebra **36** (2008), 1489–1499.

[21] V. Bardakov, V. Shpilrain and V. Tolstykh, *On the palindromic and primitive widths of a free group*, J. Algebra **285** (2005), 574–585.

[22] L. Bartholdi, *A Wilson group of non-uniformly exponential growth*, C. R. Math. Acad. Sci. Paris **336** (2003), 549–554.

[23] F. Bassino, I. Kapovich, M. Lohrey, A. Myasnikov, C. Nicaud, A. Nikolaev, I. Rivin, V. Shpilrain, A. Ushakov and P. Weil, *Complexity and Randomness in Group Theory: GAGTA Book 1*, Walter de Gruyter, 2020.

[24] F. Bassino, C. Nicaud and P. Weil, *The average-case complexity of the Word Problem for groups of matrices over $\mathbb{Z}$ is linear*, preprint, https://arxiv.org/abs/2506.00948.

[25] G. Baumslag, *Some aspects of groups with unique roots*, Acta Math. **104** (1960), 217–303.

[26] G. Baumslag, *On the residual nilpotence of certain one-relator groups*, Commun. Pure Appl. Math. **21** (1968), 491–506.

[27] G. Baumslag, *A non-cyclic one-relator group all of whose finite quotients are cyclic*, J. Aust. Math. Soc. **10** (1969), 497–498.
[28] G. Baumslag, *Positive one-relator groups*, Trans. Am. Math. Soc. **156** (1971), 165–183.
[29] G. Baumslag, *A finitely presented metabelian group with a free abelian derived group of infinite rank*, Proc. Am. Math. Soc. **35** (1972), 61–62.
[30] G. Baumslag, Some problems on one-relator groups, in: *Proceedings of the Second International Conference on the Theory of Groups (Berlin)* (Australian Nat. Univ., Canberra, 1973), Lecture Notes in Math., **372**, 75–81, Springer, 1974.
[31] G. Baumslag, Some open problems, in: *Summer School in Group Theory* (Banff, 1996). CRM Proceedings and Lecture Notes, **17**, 1–9, Amer. Math. Soc., Providence, 1999.
[32] G. Baumslag, F. Cannonito and D. Robinson, *The algorithmic theory of finitely generated metabelian groups*, Trans. Am. Math. Soc. **344** (1994), 629–648.
[33] G. Baumslag, S. M. Gersten, M. Shapiro and H. Short, Automatic groups and amalgams—a survey, in: *Algortims and Classification in Combinatorial Group Theory* (Berkeley, CA, 1989), Math. Sci. Res. Inst. Publ., **23**, 179–194, Springer, New York, 1992.
[34] G. Baumslag, C. F. Miller and G. Ostheimer, *Decomposability of finitely generated torsion-free nilpotent groups*, Int. J. Algebra Comput. **26** (2016), 1529–1546.
[35] G. Baumslag, A. Myasnikov and V. Remeslennikov, *Malnormality is decidable in free groups*, Int. J. Algebra Comput. **9** (1999), 687–692.
[36] G. Baumslag, A. Myasnikov and V. Remeslennikov, *Algebraic geometry over groups I: Algebraic sets and ideal theory*, J. Algebra **219** (1999), 16–79.
[37] G. Baumslag, A. Myasnikov and V. Roman'kov, *Two theorems about equationally Noetherian groups*, J. Algebra **194** (1997), 654–664.
[38] G. Baumslag, A. G. Myasnikov and V. Shpilrain, Open problems in combinatorial group theory, in: *Groups, Languages and Geometry* (South Hadley, MA, 1998), Contemp. Math., **250**, 1–27, Amer. Math. Soc., 1999.
[39] G. Baumslag, A. G. Myasnikov and V. Shpilrain, *Open Problems in Combinatorial Group Theory*, 2nd edn., Contemp. Math., **296**, 1–38, Amer. Math. Soc., 2002.
[40] G. Baumslag and D. Solitar, *Some two-generator one-relator non-Hopfian groups*, Bull. Am. Math. Soc. **68** (1962), 199–201.
[41] J. Behrstock and C. Drutu, *Divergence, thick groups, and short conjugators*, Ill. J. Math. **58**(4) (2014), 939–980.
[42] M. C. Bell, S. Schleimer, *The word problem for the mapping class group in quasi-linear time*, preprint, arXiv:2511.02459.
[43] N. Bergeron, *Toute variété de dimension 3 compacte et asphérique est virtuellement de Haken (d'apres Ian Agol et Daniel T. Wise)*. Astérisque No. 367–368 (2015), Exp. No. 1078, vii–viii, 115–150. ISBN 978-2-85629-804-6.
[44] G. Bergman, *Supports of derivations, free factorizations, and ranks of fixed subgroups in free groups*, Trans. Am. Math. Soc. **351** (1999), 1531–1550.
[45] M. Bestvina, *Questions in geometric group theory*, http://www.math.utah.edu/\relax \TU\texttildelow bestvina/eprints/questions-updated.pdf.
[46] M. Bestvina and N. Brady, *Morse theory and finiteness properties of groups*, Invent. Math. **129** (1997), 445–470.
[47] M. Bestvina, M. Feighn and M. Handel, *The Tits alternative for Out(F_n). I: Dynamics of exponentially growing automorphisms*, Ann. Math. **151** (2000), 517–623.
[48] M. Bestvina and M. Handel, *Train tracks and automorphisms of free groups*, Ann. Math. **135** (1992), 1–53.
[49] S. Bigelow, *The Burau representation is not faithful for n* = 5, Geom. Topol. **3** (1999), 397–404 (electronic).
[50] S. Bigelow, *Braid groups are linear*, J. Am. Math. Soc. **14** (2001), 471–486.

[51]   J. S. Birman, *Braids, Links and Mapping Class Groups*, Ann. Math. Studies, **82**, Princeton Univ. Press, 1974.
[52]   J. Birman, K. H. Ko and S. J. Lee, *A new approach to the word and conjugacy problems in the braid groups*, Adv. Math. **139** (1998), 322–353.
[53]   O. Bogopolski and O. Maslakova, *An algorithm for finding a basis of the fixed point subgroup of an automorphism of a free group*, Int. J. Algebra Comput. **26** (2016), 29–67.
[54]   W. W. Boone, *The word problem*, Ann. Math. (2) **70** (1959), 207–265.
[55]   A. Borovik, A. G. Myasnikov and V. Shpilrain, *Measuring Sets in Infinite Groups*, Contemp. Math., **298**, 21–42, Amer. Math. Soc., 2002.
[56]   A. V. Borshev and D. I. Moldavanskii [56], *On isomorphism of some groups with one defining relation*, Math. Notes **79** (2006), 31–40.
[57]   N. Brady, *Branched coverings of cubical complexes and subgroups of hyperbolic groups*, J. Lond. Math. Soc. (2) **60** (1999), 461–480.
[58]   N. Brady and J. Crisp, *CAT(0) and CAT(-1) dimensions of torsion free hyperbolic groups*, Comment. Math. Helv. **82** (2007), 61–85.
[59]   M. Bridson and D. Wise, *Malnormality is undecidable in hyperbolic groups*, Isr. J. Math. **124** (2001), 313–316.
[60]   R. Bryant, *The verbal topology of a group*, J. Algebra **48** (1977), 340–346.
[61]   M. Burger and S. Mozes, *Finitely presented simple groups and products of trees*, C. R. Acad. Sci. Paris, Sér. I Math. **324** (1997), 747–752.
[62]   M. Burger and S. Mozes, *Groups acting on trees: from local to global structure*, Publ. Math. Inst. Hautes Études Sci. **92** (2000), 113–150.
[63]   J. Burillo and E. Ventura, *Counting primitive elements in free groups*, Geom. Dedic. **93** (2002), 143–162.
[64]   R. Burns, *On the intersection of finitely generated subgroups of a free group*, Math. Z. **119** (1971), 121–130.
[65]   R. G. Burns and O. Macedonska, *Balanced presentations of the trivial group*, Bull. Lond. Math. Soc. **25** (1993), 513–526.
[66]   D. Calegari and A. Walker, *Random groups contain surface subgroups*, J. Am. Math. Soc. **28** (2015), 383–419.
[67]   M. Calvez and B. Wiest, *A fast solution to the conjugacy problem in the 4-strand braid group*, J. Group Theory **17** (2014), 757–780.
[68]   R. Camm, *Simple free products*, J. Lond. Math. Soc. **28** (1953), 66–76.
[69]   J. W. Cannon, W. J. Floyd and W. R. Parry, *Introductory notes on Richard Thompson's groups*, Enseign. Math. (2) **42** (1996), 215–256.
[70]   J. W. Cannon and W. P. Thurston, *Group invariant Peano curves*, Geom. Topol. **11** (2007), 1315–1355.
[71]   L. Chen and Y. Lodha, *The Wiegold problem and free products of left-orderable groups*, preprint, arXiv:2510.26073.
[72]   M. Chiodo, *Finding non-trivial elements and splittings in groups*, J. Algebra **331** (2011), 271–284.
[73]   L. Ciobanu, A. Levine and A. Logan, *Post's Correspondence Problem for hyperbolic and virtually nilpotent groups*, Bull. Lond. Math. Soc. **56** (2024), 159–175.
[74]   L. Ciobanu and A. Logan, *Variations on the Post correspondence problem for free groups*, Lect. Notes Comput. Sci. **12811** (2021), 90–102.
[75]   A. Clark and R. Goldstein, *Stability of numerical invariants in free groups*, Commun. Algebra **33** (2005), 4097–4104.
[76]   A. Clifford and R. Z. Goldstein, *Tesselations of S^2 and equations over torsion-free groups*, Proc. Edinb. Math. Soc. (2) **38** (1995), 485–493.
[77]   A. Clifford and R. Goldstein, *Subgroups of free groups and primitive elements*, J. Group Theory **13** (2010), 601–611.
[78]   M. Cohen and M. Lustig, *On the dynamics and the fixed subgroup of a free group automorphism*, Invent. Math. **96** (1989), 613–638.

[79] M. M. Cohen and M. Lustig, *The conjugacy problem for Dehn twist automorphisms of free groups*, Comment. Math. Helv. **74** (1999), 179–200.

[80] M. Cohen, W. Metzler and A. Zimmermann, *What does a basis of F(a, b) look like?*, Math. Ann. **257** (1981), 435–445.

[81] D. Collins, M. Edjvet and C. P. Gill, *Growth series for the group $\langle x, y | x^{-1}yx = y^l \rangle$*, Arch. Math. (Basel) **62** (1994), 1–11.

[82] D. Collins and E. C. Turner, *All automorphisms of free groups with maximal rank fixed subgroups*, Math. Proc. Camb. Philos. Soc. **119** (1996), 615–630.

[83] G. Conner, *Central extensions of word hyperbolic groups satisfy a quadratic isoperimetric inequality*, Arch. Math. **65** (1995), 465–479.

[84] A. R. Conway and A. J. Gutmann, *Square lattice self-avoiding walks and polygons*, Ann. Comb. **5** (2001), 319–345.

[85] D. Cooper, *Automorphisms of free groups have finitely generated fixed point sets*, J. Algebra **111** (1987), 453–456.

[86] T. Coulbois and A. Khelif, *Equations in free groups are not finitely approximable*, Proc. Am. Math. Soc. **127** (1999), 963–965.

[87] R. Craggs, *Free Heegaard diagrams and extended Nielsen transformations. I*, Mich. Math. J. **26** (1979), 161–186.

[88] R. Craggs, *Free Heegaard diagrams and extended Nielsen transformations. II*, Ill. J. Math. **23** (1979), 101–127.

[89] F. Dahmani and V. Guirardel, *The isomorphism problem for all hyperbolic groups*, Geom. Funct. Anal. **21** (2011), 223–300.

[90] A. Darbinyan, *Computable groups which do not embed into groups with decidable conjugacy problem*, Invent. Math. **224** (2021), 987–997.

[91] P. de la Harpe, *Topics in Geometric Group Theory*, University of Chicago Press, 2000.

[92] P. de la Harpe, *Uniform growth in groups of exponential growth*, Geom. Dedic. **95** (2002), 1–17.

[93] M. Dehn, *Über unendliche diskontinuierliche Gruppen*, Math. Ann. **71**(1) (1911), 116–144 (German).

[94] P. Dehornoy, *A fast method for comparing braids*, Adv. Math. **125** (1997), 200–235.

[95] P. Dehornoy, *Braids and Self-Distributivity*, Progress in Mathematics, **192**, Birkhäuser Verlag, Basel, 2000.

[96] T. Delzant, *Group rings of hyperbolic groups*, C. R. Acad. Sci. Paris, Sér. I Math. **324** (1997), 381–384.

[97] W. Dicks, *Equivalence of the strengthened Hanna Neumann conjecture and the amalgamated graph conjecture*, Invent. Math. **117** (1994), 373–389.

[98] W. Dicks and E. Formanek, *The rank three case of the Hanna Neumann conjecture*, J. Group Theory **4** (2001), 113–151.

[99] W. Dicks and I. Leary, *Presentations for subgroups of Artin groups*, Proc. Am. Math. Soc. **127** (1999), 343–348.

[100] W. Dicks and E. Ventura, *The Group Fixed by a Family of Injective Endomorphisms of a Free Group*, Contemporary Math., **195**, Amer. Math. Soc., 1996.

[101] E. Dinowitz, L. Koch-Hyde, S. O'Connor and E. Olive, *Growth of potentially positive elements in the free group on two generators*, preprint.

[102] C. Droms, J. Lewin and H. Servatius, *The length of elements in free solvable groups*, Proc. Am. Math. Soc. **119** (1993), 27–33.

[103] M. Dunwoody, *On verbal subgroups of free groups*, Arch. Math. **16** (1965), 153–157.

[104] V. G. Durnev, *The Mal'tsev–Nielsen equation on a free metabelian group of rank 2*, Math. Notes USSR **46** (1989), 927–929.

[105] V. G. Durnev, *A generalization of Problem 9.25 in the Kourovka notebook*, Math. Notes USSR **47** (1990), 117–121.

[106] J. Dyer, E. Formanek and E. Grossman, *On the linearity of automorphism groups of free groups*, Arch. Math. **38** (1982), 404–409.

[107] J. Dyer and E. Grossman, *The automorphism groups of the braid groups*, Am. J. Math. **103** (1981), 1151–1169.

[108] M. Edjvet and D. Johnson, *The growth of certain amalgamated free products and HNN extensions*, J. Aust. Math. Soc. A **52** (1993), 285–298.

[109] D. Epstein, J. Cannon, D. Holt, S. Levy, M. Paterson and W. Thurston, *Word Processing in Groups*, Jones and Bartlett Publishers, Boston, 1992.

[110] D. Epstein and D. Holt, *The linearity of the conjugacy problem in word-hyperbolic groups*, Int. J. Algebra Comput. **16**(2) (2006), 287–305.

[111] A. Eskin, S. Mozes and H. Oh, *Uniform exponential growth for linear groups*, Int. Math. Res. Not. **31** (2002), 1675–1683.

[112] M. Evans, *Primitive elements in the free metabelian group of rank 3*, J. Algebra **220** (1999), 475–491.

[113] M. Feighn and M. Handel, *Mapping tori of free group automorphisms are coherent*, Ann. Math. **149** (1999), 1061–1077.

[114] E. Formanek, *Fixed points and centers of automorphism groups of free nilpotent groups*, Commun. Algebra **30** (2002), 1033–1038.

[115] E. Formanek and C. Procesi, *The automorphism group of a free group is not linear*, J. Algebra **149** (1992), 494–499.

[116] N. Franco and J. González-Meneses, *Conjugacy problem for braid groups and Garside groups*, J. Algebra **266** (2003), 112–132.

[117] J. Friedman, *Sheaves on graphs, their homological invariants, and a proof of the Hanna Neumann conjecture: with an appendix by Warren Dicks*, Mem. Am. Math. Soc. **233** (2015), 1100.

[118] A. M. Gaglione, A. G. Myasnikov, V. N. Remeslennikov and D. Spellman, *Formal power series representations of free exponential groups*, Commun. Algebra **25** (1997), 631–648.

[119] G. Gardam, *A counterexample to the unit conjecture for group rings*, Ann. Math. (2) **194** (2021), 967–979.

[120] G. Gardam, *Non-trivial units of complex group rings*, preprint, arXiv:2312.05240.

[121] A. Garreta, A. Myasnikov and D. Ovchinnikov, *Diophantine problems in solvable groups*, Bull. Math. Sci. **10**(1) (2020), 2050005.

[122] F. A. Garside, *The braid group and other groups*, Q. J. Math. Oxford Ser. (2) **20** (1969), 235–254.

[123] V. Gebhardt, *A new approach to the conjugacy problem in Garside groups*, J. Algebra **292** (2005), 282–302.

[124] S. Gersten, *On fixed points of automorphisms of finitely generated free groups*, Bull. Am. Math. Soc. **8** (1983), 451–454.

[125] S. Gersten, *Fixed points of automorphisms of free groups*, Adv. Math. **64** (1987), 51–85.

[126] S. Gersten, *Isodiametric and isoperimetric inequalities in group extensions*, preprint, University of Utah, 1991.

[127] S. M. Gersten, Problems on automatic groups, in: *Algorithms and Classification in Combinatorial Group Theory* (Berkeley, CA, 1989), Math. Sci. Res. Inst. Publ., **23**, 225–232, Springer, New York, 1992.

[128] S. M. Gersten, Dehn functions and l_1-norms of finite presentations. in: *Algorithms and Classification in Combinatorial Group Theory* (Berkeley, CA, 1989), Math. Sci. Res. Inst. Publ., **23**, 195–224, Springer, New York, 1992.

[129] S. M. Gersten, D. F. Holt and T. R. Riley, *Isoperimetric inequalities for nilpotent groups*, Geom. Funct. Anal. **13** (2003), 795–814.

[130] D. Gildenhuys, O. Kharlampovich and A. Myasnikov, *CSA groups and separated free constructions*, Bull. Aust. Math. Soc. **52** (1995), 63–84.

[131] R. Gitik, M. Mitra, E. Rips and M. Sageev, *Widths of subgroups*, Trans. Am. Math. Soc. **350** (1998), 321–329.

[132] R. Goebel and A. Paras, *Outer automorphism groups of metabelian groups*, J. Pure Appl. Algebra **149** (2000), 251–266.

[133] R. Goldstein, *An algorithm for potentially positive words in F_2*, Contemp. Math. **421** (2006), 157–168.

[134] R. Goldstein and E. Turner, *Fixed subgroups of homomorphisms of free groups*, Bull. Lond. Math. Soc. **18** (1986), 468–470.

[135] E. Golod, *On nil-algebras and finitely approximable p-groups*, Izv. Akad. Nauk SSSR, Ser. Mat. **28** (1964), 273–276 (Russian).

[136] J. Gonzalez-Meneses, *The n-th root of a braid is unique up to conjugacy*, Algebr. Geom. Topol. **3** (2003), 1103–1118.

[137] R. D. Gray, *Undecidability of the word problem for one-relator inverse monoids via right-angled Artin subgroups of one-relator groups*, Invent. Math. **219** (2020), 987–1008.

[138] R. I. Grigorchuk, *On the Milnor problem of group growth*, Dokl. Akad. Nauk SSSR **271** (1983), 30–33 (Russian).

[139] R. I. Grigorchuk, *Construction of p-groups of intermediate growth that have a continuum of factor-groups*, Algebra Log. **23** (1984), 383–394, 478 (Russian).

[140] R. I. Grigorchuk, *Degrees of growth of p-groups and torsion-free groups*, Mat. Sb. **126**(168) (1985), 194–214, 286 (Russian).

[141] R. I. Grigorchuk and P. de la Harpe, *On problems related to growth, entropy and spectrum in group theory*, J. Dyn. Control Syst. **3** (1997), 51–89.

[142] R. I. Grigorchuk and P. de la Harpe, *One-relator groups of exponential growth have uniformly exponential growth*, Math. Notes **69** (2001), 628–630.

[143] R. Grigorchuk and P. Kurchanov, *On the width of elements in free groups*, Ukr. Mat. Zh. **43** (1991), 911–918.

[144] R. I. Grigorchuk and P. F. Kurchanov, Some questions of group theory related to geometry, in: *Algebra VII*, Encyclopaedia Math. Sci., **58**, 167–240, Springer, Berlin, 1993. Translated from the Russian by P. M. Cohn.

[145] R. I. Grigorchuk and A. Maki, *On a group of intermediate growth that acts on a line by homeomorphisms*, Mat. Zametki **53** (1993), 46–63 (Russian); translation in Math. Notes **53** (1993), 146–157.

[146] M. Gromov, Hyperbolic groups, in: *Essays in Group Theory*, Math. Sci. Res. Inst. Publ., **8**, 75–263, Springer, New York, 1987.

[147] M. Gromov, Asymptotic invariants of infinite groups, in: *Geometric Group Theory, Vol. 2* (Sussex, 1991), London Math. Soc. Lecture Note Ser., **182**, 1–295, Cambridge Univ. Press, Cambridge, 1993.

[148] J. R. J. Groves and C. F. Miller III, *Recognizing free metabelian groups*, Ill. J. Math. **30** (1986), 246–254.

[149] V. S. Guba, *Equivalence of infinite systems of equations in free groups and semigroups to finite subsystems*, Math. Notes USSR **40** (1986), 688–690.

[150] V. S. Guba, *On the Ore condition for the group ring of R. Thompson's group F*, Commun. Algebra **49** (2021), 4699–4711.

[151] V. Guba, *Amenability problem for Thompson's group F: state of the art*, J. Groups Complex. Cryptol. **15**(1) (2023), 3.

[152] C. K. Gupta and F. Levin, *Tame range of automorphism groups of free polynilpotent groups*, Commun. Algebra **19** (1991), 2497–2500.

[153] C. K. Gupta and V. Shpilrain, *The centre of a one-relator solvable group*, Int. J. Algebra Comput. **3** (1993), 51–55.

[154] N. Gupta and V. Shpilrain, *Nielsen's commutator test for two-generator groups*, Math. Proc. Camb. Philos. Soc. **114** (1993), 295–301.

[155] C. K. Gupta and V. Shpilrain, Lifting automorphisms: a survey, in: *Groups '93 Galway/St. Andrews, Vol. 1* (Galway, 1993), London Math. Soc. Lecture Note Ser., **211**, 249–263, Cambridge Univ. Press, Cambridge, 1995.

[156] F. Haglund and D. T. Wise, *Special cube complexes*, Geom. Funct. Anal. **17** (2008), 1551–1620.

[157] J. Hamkins, *Every group has a terminating transfinite automorphism tower*, Proc. Am. Math. Soc. **126** (1998), 3223–3226.

[158] J. Harlander, *Solvable groups with cyclic relation module*, J. Pure Appl. Algebra **90** (1993), 189–198.

[159] D. Harvey and J. van der Hoeven, *Integer multiplication in time O(n log n)*, Ann. Math. **193** (2021), 563–617.
[160] G. Havas and M. Vaughan-Lee, *4-Engel groups are locally nilpotent*, Int. J. Algebra Comput. **15** (2005), 649–682.
[161] H. Heineken, *Engelsche elemente der länge drei*, Ill. J. Math. **5** (1961), 681–707 (German).
[162] J. Helfer and D. T. Wise, *Counting cycles in labeled graphs: the nonpositive immersion property for one-relator groups*, Int. Math. Res. Not. **9** (2016), 2813–2827.
[163] M. Heusener and R. Weidmann, *A remark on Whitehead's cut-vertex lemma*, J. Group Theory **22** (2019), 15–21.
[164] C. Hidber, *Isoperimetric functions of finitely generated nilpotent groups*, J. Pure Appl. Algebra **144** (1999), 229–242.
[165] R. Hirshon, *Some properties of endomorphisms in residually finite groups*, J. Aust. Math. Soc. A **24** (1977), 117–120.
[166] W. Hodges, *Model Theory*, Encyclopedia of Mathematics and its Applications, **42**, Cambridge University Press, Cambridge, 1993, xiv+772 pp.
[167] C. Hog-Angeloni and W. Metzler, The Andrews–Curtis conjecture and its generalizations, in: *Two-dimensional Homotopy and Combinatorial Group Theory*, London Math. Soc. Lecture Note Ser., **197**, 365–380, Cambridge Univ. Press, Cambridge, 1993.
[168] D. Holt, M. Lohrey and S. Schleimer, *Compressed decision problems in hyperbolic groups*, Groups Geom. Dyn. **18** (2024), 1233–1273.
[169] R. Horowitz, *Characters of free groups represented in the two-dimensional special linear group*, Commun. Pure Appl. Math. **25** (1972), 635–649.
[170] J. Howie, *Some remarks on a problem of J. H. C. Whitehead*, Topology **22** (1983), 475–485.
[171] J. Howie, How to generalize one-relator group theory, in: *Combinatorial Group Theory and Topology* (Alta, Utah, 1984), Ann. of Math. Stud., **111**, 53–78, Princeton Univ. Press, 1987.
[172] J. Howie, *A proof of the Scott–Wiegold conjecture on free products of cyclic groups*, J. Pure Appl. Algebra **173** (2002), 167–176.
[173] J. Huebschmann, *Aspherical 2-complexes and an unsettled problem of J. H. C. Whitehead*, Math. Ann. **258** (1981/82), 17–37.
[174] S. Humphries, *Torsion-free quotients of braid groups*, Int. J. Algebra Comput. **11** (2001), 363–373.
[175] S. Humphries, *Braid groups and Aut(F_2) are not rigid*, Contemp. Math. **360** (2004), 51–54.
[176] S. Humphries, *Representations and rigidity of Aut(F_3)*, Int. J. Algebra Comput. **16** (2006), 925–929.
[177] L. Hyde, S. O'Connor and V. Shpilrain, *Orbit-blocking words and the average-case complexity of Whitehead's problem in the free group of rank 2*, J. Group Theory **28** (2025), 423–430.
[178] W. Imrich and E. C. Turner, *Endomorphisms of free groups and their fixed points*, Math. Proc. Camb. Philos. Soc. **105** (1989), 421–422.
[179] S. V. Ivanov, *The free Burnside groups of sufficiently large exponents*, Int. J. Algebra Comput. **4** (1994), 01n02, 308 pp.
[180] S. Ivanov, *On certain elements of free groups*, J. Algebra **204** (1998), 394–405.
[181] S. Ivanov, *On endomorphisms of a free group that preserve primitivity*, Arch. Math. **72** (1999), 92–100.
[182] S. V. Ivanov, *An asphericity conjecture and Kaplansky problem on zero divisors*, J. Algebra **216** (1999), 13–19.
[183] S. V. Ivanov, *On Rourke's extension of group presentations and a cyclic version of the Andrews--Curtis conjecture*, Proc. Am. Math. Soc. **134** (2006), 1561–1567.
[184] S. V. Ivanov, *On balanced presentations of the trivial group*, Invent. Math. **165** (2006), 525–549.
[185] S. V. Ivanov and A. Yu. Ol'shanskii, *Hyperbolic groups and their quotients of bounded exponents*, Trans. Am. Math. Soc. **348**(6) (1996), 2091–2138.
[186] S. V. Ivanov and P. Schupp, A remark on finitely generated subgroups of free groups, in: *Algorithmic Problems in Groups and Semigroups*, 139–142, Birkhauser, 2000.

[187] A. Jaikin-Zapirain, *Free ℚ-groups are residually torsion-free nilpotent*, Ann. Sci. Éc. Norm. Supér. (4) **57** (2024), 1101–1133.

[188] A. Jaikin-Zapirain and M. Linton, *On the coherence of one-relator groups and their group algebras*, Ann. Math. (2) **201** (2025), 909–959.

[189] J. M. T. Jones, *Direct products and the Hopf property*, J. Aust. Math. Soc. **17** (1974), 174–196.

[190] J. M. T. Jones, On isomorphisms of direct powers, in: *Word Problems, II (Conf. on Decision Problems in Algebra, Oxford, 1976)*, Stud. Logic Foundations Math., **95**, 215–245, North-Holland, 1980.

[191] M. Kapovich, *Representations of polygons of finite groups*, Geom. Topol. **9** (2005), 1915–1951.

[192] I. Kapovich and N. Benakli, in: Boundaries of hyperbolic groups, *Combinatorial And Geometric Group Theory* (New York, 2000/Hoboken, NJ, 2001), Contemp. Math., **296**, 39–93, Amer. Math. Soc., Providence, RI, 2002.

[193] I. Kapovich, G. Levitt, P. Schupp and V. Shpilrain, *Translation equivalence in free groups*, Trans. Am. Math. Soc. **359** (2007), 1527–1546.

[194] I. Kapovich, A. Miasnikov, P. Schupp and V. Shpilrain, *Average-case complexity and decision problems in group theory*, Adv. Math. **190** (2005), 343–359.

[195] I. Kapovich and A. G. Myasnikov, *Stallings foldings and subgroups of free groups*, J. Algebra **248** (2002), 608–668.

[196] I. Kapovich, A. G. Myasnikov, P. Schupp and V. Shpilrain, *Generic-case complexity, decision problems in group theory and random walks*, J. Algebra **264** (2003), 665–694.

[197] I. Kapovich, P. Schupp and V. Shpilrain, *Generic properties of Whitehead's algorithm and isomorphism rigidity of random one-relator groups*, Pac. J. Math. **223** (2006), 113–140.

[198] I. Kapovich and D. Wise, *The equivalence of some residual properties of word-hyperbolic groups*, J. Algebra **223** (2000), 562–583.

[199] A. Karrass and D. Solitar, *On finitely generated subgroups of a free group*, Proc. Am. Math. Soc. **22** (1969), 209–213.

[200] M. Kassabov, *An automorphism of a free metabelian group without fixed points*, Commun. Algebra **32** (2004), 3297–3303.

[201] B. Khan, Positively generated subgroups of free groups and the Hanna Neumann conjecture, in: *Combinatorial and Geometric Group Theory* (New York, 2000/Hoboken, NJ, 2001), Contemp. Math., **296**, 155–170, Amer. Math. Soc., Providence, RI, 2002.

[202] B. Khan, *The structure of automorphic conjugacy in the free group of rank 2*, Contemp. Math. **349** (2004), 115–196.

[203] O. G. Kharlampovich, *A finitely presented solvable group with unsolvable word problem*, Izv. Akad. Nauk SSSR, Ser. Mat. **45** (1981), 852–873 (Russian).

[204] O. Kharlampovich, I. G. Lysenok, A. G. Miasnikov and N. W. M. Touikan, *Quadratic equations over free groups are NP-complete*, Theory Comput. Syst. **45** (2009), 1432–4350.

[205] O. Kharlampovich and A. Myasnikov, *Irreducible affine varieties over groups. I, II*, J. Algebra **200** (1998), 472–516, 517–570.

[206] O. Kharlampovich and A. Myasnikov, *Hyperbolic groups and free constructions*, Trans. Am. Math. Soc. **350** (1998), 571–613.

[207] O. Kharlampovich and A. Myasnikov, *Tarski's problem about the elementary theory of free groups has a positive solution*, Electron. Res. Announc. Am. Math. Soc. **4** (1998), 101–108 (electronic).

[208] O. Kharlampovich and A. Myasnikov, Effective JSJ decompositions, in: *Groups, Languages, Algorithms*, Contemp. Math., **378**, 87–212, Amer. Math. Soc., Providence, RI, 2005.

[209] O. Kharlampovich and A. Myasnikov, Algebraic geometry over free groups: lifting solutions into generic points, in: *Groups, Languages, Algorithms*, Contemp. Math., **378**, 213–318, Amer. Math. Soc., Providence, RI, 2005.

[210] O. Kharlampovich and A. Myasnikov, *Implicit function theorem over free groups*, J. Algebra **290** (2005), 1–203.

[211] O. Kharlampovich and A. Myasnikov, *Elementary theory of free nonabelian groups*, J. Algebra **302** (2006), 451–552.

[212] O. Kharlampovich and A. Myasnikov, *Definable sets in a hyperbolic group*, Int. J. Algebra Comput. **23** (2013), 91–110.

[213] O. Kharlampovich and A. Myasnikov, Model theory and algebraic geometry in groups, non-standard actions and algorithmic problems, in: *Proceedings of the International Congress of Mathematicians—Seoul 2014. Vol. II*, 223–245, 2014.

[214] D. G. Khramtsov, *Finite groups of automorphisms of free groups*, Math. Notes **38** (1985), 721–724.

[215] A. A. Klyachko, *A funny property of sphere and equations over groups*, Commun. Algebra **21** (1993), 2555–2575.

[216] D. Knuth, J. H. Morris and V. Pratt, *Fast pattern matching in strings*, SIAM J. Comput. **6**(2) (1977), 323–350.

[217] L. Koch-Hyde, S. O'Connor and E. Olive, *A new algorithm for determining potential positivity in the free group of rank 2*, Int. J. Alg. Comput., to appear.

[218] L. Koch-Hyde, S. O'Connor, E. Olive and V. Shpilrain, *Orbit-blocking words in free groups*, J. Group Theory, to appear. arXiv:2505.00477.

[219] K. Kordek and D. Margalit, *Homomorphisms of commutator subgroups of braid groups*, Bull. Lond. Math. Soc. **54** (2022), 95–111.

[220] A. I. Kostrikin, *Around Burnside*, Springer Series of Modern Surveys in Mathematics, **20**, 1990.

[221] M. Koubi, *Croissance uniforme dans les groupes hyperboliques*, Ann. Inst. Fourier **48** (1998), 1441–1453.

[222] D. Krammer, *The braid group B_4 is linear*, Invent. Math. **142** (2000), 451–486.

[223] D. Krammer, *Braid groups are linear*, Ann. Math. (2) **155** (2002), 131–156.

[224] R. Kropholler, Special cube complexes, in: *Geometric and Cohomological Group Theory*, London Math. Soc. Lecture Note Ser., **444**, 46–66, Cambridge Univ. Press, Cambridge, 2018.

[225] P. Kropholler, P. Linnell and J. Moody, *Applications of a new K-theoretic theorem to soluble group rings*, Proc. Am. Math. Soc. **104** (1988), 675–684.

[226] Yu.V. Kuz'min, *Homology theory of free abelianized extensions*, Commun. Algebra **16** (1988), 2447–2533.

[227] I. Leary, *Asphericity and zero divisors in group algebras*, J. Algebra **227** (2000), 362–364.

[228] D. Lee, *On certain C-test words for free groups*, J. Algebra **247** (2002), 509–540.

[229] D. Lee, *Primitivity preserving endomorphisms of free groups*, Commun. Algebra **30** (2002), 1921–1947.

[230] D. Lee, *Counting words of minimum length in an automorphic orbit*, J. Algebra **301** (2006), 35–58.

[231] D. Lee, *A tighter bound for the number of words of minimum length in an automorphic orbit*, J. Algebra **305** (2006), 1093–1101.

[232] D. Lee, *Translation equivalent elements in free groups*, J. Group Theory **9** (2006), 809–814.

[233] D. Lee, *An algorithm that decides translation equivalence in a free group of rank two*, J. Group Theory **10** (2007), 561–569.

[234] D. Lee, *On several problems about automorphisms of the free group of rank two*, J. Algebra **321** (2009), 167–193.

[235] G. Levitt and M. Lustig, *Periodic ends, growth rates, Hölder dynamics for automorphisms of free groups*, Comment. Math. Helv. **75** (2000), 415–429.

[236] V. Lin, *Braids and permutations*, preprint, arXiv:math/0404528.

[237] P. Linnell and T. Schick, *Finite group extensions and the Atiyah conjecture*, J. Am. Math. Soc. **20** (2007), 1003–1051.

[238] M. Linton, *Hyperbolic one-relator groups*, Can. J. Math. Published online, May 2024. https://doi.org/10.4153/S0008414X24000427.

[239] M. Linton, *One-relator hierarchies*, Duke Math. J. **174** (2025), 747–802.

[240] M. Linton and C.-F. Nyberg-Brodda, *The theory of one-relator groups: history and recent progress*, preprint, 2025, arXiv:2501.18306.

[241] S. Lipschutz, *Generalization of Dehn's result on the conjugacy problem*, Proc. Am. Math. Soc. **17** (1966), 759–762.

[242] S. Lipschutz, *The conjugacy problem and cyclic amalgamations*, Bull. Am. Math. Soc. **81** (1975), 114–116.
[243] M. Lohrey, *Word problems and membership problems on compressed words*, SIAM J. Comput. **35** (2006), 1210–1240.
[244] M. Lohrey, *The Compressed Word Problem for Groups*, SpringerBriefs in Mathematics, Springer, New York, 2014, xii+153 pp.
[245] M. Lohrey, *Knapsack in hyperbolic groups*, J. Algebra **545** (2020), 390–415.
[246] M. Lohrey, *Subgroup membership in $GL(2,\mathbb{Z})$*, Theory Comput. Syst. **68** (2024), 1082–1107.
[247] D. Long and M. Paton, *The Burau representation is not faithful for $n \geq 6$*, Topology **32** (1993), 439–447.
[248] J. Los, *On the conjugacy problem for automorphisms of free groups*, Topology **35** (1996), 779–808.
[249] L. Louder and H. Wilton, *Stackings and the W-cycles conjecture*, Can. Math. Bull. **60** (2017), 604–612.
[250] L. Louder and H. Wilton, *One-relator groups with torsion are coherent*, Math. Res. Lett. **27** (2020), 1499–1512.
[251] L. Louder and H. Wilton, *Negative immersions for one-relator groups*, Duke Math. J. **171** (2022), 547–594.
[252] L. Louder and H. Wilton, *Uniform negative immersions and the coherence of one-relator groups*, Invent. Math. **236** (2024), 673–712.
[253] E. Luft, *On 2-dimensional aspherical complexes and a problem of J. H. C. Whitehead*, Math. Proc. Camb. Philos. Soc. **119** (1996), 493–495.
[254] R. Lyndon, *Groups with parametric exponents*, Trans. Am. Math. Soc. **96** (1960), 518–533.
[255] R. Lyndon, Problems in combinatorial group theory, in: *Combinatorial Group Theory and Topology* (Alta, Utah, 1984), Ann. of Math. Stud., **111**, 3–33, Princeton Univ. Press, 1987.
[256] R. Lyndon and P. Schupp, *Combinatorial Group Theory*, Classics in Mathematics, Springer-Verlag, Berlin, 2001 (Reprint of the 1977 edition).
[257] I. G. Lysenok, *Some algorithmic properties of hyperbolic groups*, Math. USSR, Izv. **35** (1990), 145–163.
[258] I. G. Lysenok, *Infinite Burnside groups of even period*, Izv. Ross. Akad. Nauk, Ser. Mat. **60**(3) (1996), 3–224 (Russian).
[259] I. G. Lysenok and A. Ushakov, *Spherical quadratic equations in free metabelian groups*, Proc. Am. Math. Soc. **144** (2016), 1383–1390.
[260] I. G. Lysenok and A. Ushakov, *Orientable quadratic equations in free metabelian groups*, J. Algebra **581** (2021), 303–326.
[261] N. Madras and G. Slade, *The Self-avoiding Walk*, Modern Birkhäuser Classics, Birkhäuser/Springer, New York, 2013. Reprint of the 1993 original.
[262] W. Magnus, *Über discontinuierliche gruppen mit einer definierenden Relation (Der Freiheitssatz)*, J. Reine Angew. Math. **163** (1930), 141–165.
[263] W. Magnus, *Das Identitätsproblem für Gruppen mit einer definierenden Relation*, Math. Ann. **106**(1) (1932), 295–307 (German).
[264] W. Magnus, A. Karrass and D. Solitar, *Combinatorial Group Theory. Presentations of Groups in Terms of Generators and Relations*, Dover Publications, Inc., Mineola, NY, 2004. Reprint of the 1976 second edition.
[265] W. Magnus and A. Peluso, *On a theorem of V. I. Arnold*, Commun. Pure Appl. Math. **22** (1969), 683–692.
[266] G. Makanin, *Equations in a free group*, Math. USSR, Izv. **21**(3) (1983), 546–582.
[267] G. Makanin, *Decidability of the universal and positive theories of a free group*, Math. USSR, Izv. **25**(1) (1985), 75–88.
[268] J. Makowsky, *On some conjectures connected with complete sentences*, Fundam. Math. **81** (1974), 193–202.
[269] R. Mandel and A. Ushakov, *Quadratic equations in metabelian Baumslag-Solitar groups*, Int. J. Algebra Comput. **33** (2023), 1195–1216.
[270] A. Martino, *Intersections of automorphism fixed subgroups in the free group of rank three*, Algebr. Geom. Topol. **4** (2004), 177–198.

[271] A. Martino and E. Ventura, *On automorphism-fixed subgroups of a free group*, J. Algebra **230** (2000), 596–607.
[272] A. Martino and E. Ventura, *Fixed subgroups are compressed in free groups*, Commun. Algebra **32** (2004), 3921–3935.
[273] A. Martino and E. Ventura, *A description of auto-fixed subgroups of a free group*, Topology **43** (2004), 1133–1164.
[274] E. Martínez-Pedroza and D. Wise, *Local quasiconvexity of groups acting on small cancellation complexes*, J. Pure Appl. Algebra **215** (2011), 2396–2405.
[275] O. Maslakova, *The fixed point group of an automorphism of a free group*, Algebra Log. **42** (2003), 237–265.
[276] H. A. Masur and Y. N. Minsky, *Geometry of the complex of curves. II. Hierarchical structure*, Geom. Funct. Anal. **10**(4) (2000), 902–974.
[277] Yu.V. Matiyasevich, Hilbert's tenth problem: Diophantine equations in the Twentieth Century, in: *Mat. Events of the Twentieth Century*, 185–213, Springer and PHASIS, 2006.
[278] J. P. McCammond and D. T. Wise, *Coherence, local quasiconvexity, and the perimeter of 2-complexes*, Geom. Funct. Anal. **15** (2005), 859–927.
[279] J. McCool, *A characterization of periodic automorphisms of a free group*, Trans. Am. Math. Soc. **260** (1980), 309–318.
[280] J. McCool, *Free group roots of $a^k b^l$ and $[a^k, b]$*, Int. J. Algebra Comput. **10** (2000), 339–347.
[281] J. McCool, *On a question of Remeslennikov*, Glasg. Math. J. **43** (2001), 123–124.
[282] J. McCool and S. Krstic, *The non-finite presentability of $IA(F_3)$ and $GL_2(Z[t, t^{-1}])$*, Invent. Math. **129** (1997), 595–606.
[283] J. McCool and A. Pietrowski, *On free products with amalgamation of two infinite cyclic groups*, J. Algebra **18** (1971), 377–383.
[284] J. Meakin and P. Weil, *Subgroups of free groups: a contribution to the Hanna Neumann conjecture*, Geom. Dedic. **94** (2002), 33–43.
[285] Yu. Merzlyakov, *Positive formulae on free groups*, Algebra Log. **4** (1966), 25–42 (Russian).
[286] C. F. Miller III, *On Group-theoretic Decision Problems and Their Classification*, Annals of Math. Studies, **111**, 31, Princeton Univ. Press, 1987.
[287] C. F. Miller and P. Schupp, *Some Presentations of the Trivial Group*, Contemp. Math., **250**, 113–115, Amer. Math. Soc., Providence, RI, 1999.
[288] I. Mineyev, *Submultiplicativity and the Hanna Neumann conjecture*, Ann. Math. (2) **175** (2012), 393–414.
[289] M. Mitra, *Cannon–Thurston maps for hyperbolic group extensions*, Topology **37** (1998), 527–538.
[290] M. Mitra, *Cannon–Thurston maps for trees of hyperbolic metric spaces*, J. Differ. Geom. **48** (1998), 135–164.
[291] M. Mitra, Coarse extrinsic geometry: a survey, in: *The Epstein Birthday Schrift*, Geom. Topol. Monogr., **1**, 341–364, Geometry & Topology Publications, Coventry, 1998.
[292] J. Moody, *The faithfulness question for the Burau representation*, Proc. Am. Math. Soc. **119** (1993), 671–679.
[293] R. Muchnik and I. Pak, *Percolation on Grigorchuk groups*, Commun. Algebra **29** (2001), 661–671.
[294] A. Murray, *More counterexamples to the unit conjecture for group rings*, preprint, arXiv:2106.02147.
[295] A. G. Myasnikov, *Extended Nielsen transformations and the trivial group*, Mat. Zametki **35** (1984), 491–495 (Russian).
[296] A. D. Myasnikov and A. G. Myasnikov, Balanced presentations of the trivial group on two generators and the Andrews–Curtis conjecture, in: *Groups and Computation, III* (Columbus, OH, 1999), Ohio State Univ. Math. Res. Inst. Publ., **8**, 257–263, de Gruyter, Berlin, 2001.
[297] A. D. Myasnikov, A. G. Myasnikov and V. Shpilrain, On the Andrews–Curtis equivalence, in: *Contemp. Math.*, **296**, 183–198, Amer. Math. Soc., 2002.
[298] A. G. Myasnikov, A. Nikolaev and A. Ushakov, *The Post correspondence problem in groups*, J. Group Theory **17** (2014), 991–1008.

[299] A. G. Myasnikov, A. Nikolaev and A. Ushakov, *Knapsack problems in groups*, Math. Comput. **84**(292) (2015), 987–1016.

[300] A. G. Myasnikov and V. Remeslennikov, *Exponential groups. II. Extensions of centralizers and tensor completion of CSA-groups*, Int. J. Algebra Comput. **6** (1996), 687–711.

[301] A. G. Myasnikov, V. Roman'kov, A. Ushakov and A. Vershik, *The word and geodesic problems in free solvable groups*, Trans. Am. Math. Soc. **362** (2010), 4655–4682.

[302] A. G. Myasnikov and V. Shpilrain, *Automorphic orbits in free groups*, J. Algebra **269** (2003), 18–27.

[303] V. Nekrashevych, *Palindromic subshifts and simple periodic groups of intermediate growth*, Ann. Math. (2) **187** (2018), 667–719.

[304] V. Nekrashevych and S. Sidki, Automorphisms of the binary tree: state-closed subgroups and dynamics of 1/2-endomorphisms, in: *Groups: Topological, Combinatorial and Arithmetic Aspects*, LMS Lecture Notes Series, **311**, 375–404, 2004.

[305] H. Neumann, *Varieties of Groups*, Springer, 1967.

[306] W. Neumann, On intersections of finitely generated subgroups of free groups, in: *Groups—Canberra 1989*, Lecture Notes Math., **1456**, 161–170, Springer, Berlin, 1990.

[307] L. P. Neuwirth, *Knot Groups*, Annals of Mathematics Studies, **56**, Princeton University Press, Princeton, N.J., 1965.

[308] B. B. Newman, *Some results on one-relator groups*, Bull. Am. Math. Soc. **74** (1968), 568–571.

[309] G. A. Noskov, *The genus of a free metabelian group* (Russian). Preprint 84-509. Akad. Nauk SSSR Sibirsk. Otdel., Vychisl. Tsentr, Novosibirsk, 1984, 18 pp.

[310] P. S. Novikov and S. I. Adian, *Infinite periodic groups. I, II, III*, Izv. Akad. Nauk SSSR, Ser. Mat. **32** (1968), 212–244, 251–524, 709–731 (Russian).

[311] Y. Ollivier, *Sharp phase transition theorems for hyperbolicity of random groups*, Geom. Funct. Anal. **14**(3) (2004), 595–679.

[312] A. Olshanskii, *The finite basis problem for identities in groups*, Izv. Akad. Nauk SSSR, Ser. Mat. **34** (1970), 376–384.

[313] A. Yu. Ol'shanskii, *Geometry of Defining Relations in Groups*, Mathematics and its Applications (Soviet Series), **70**, Kluwer Academic Publishers Group, Dordrecht, 1991.

[314] A. Yu. Ol'shanskii, *On residualing homomorphisms and G-subgroups of hyperbolic groups*, Int. J. Algebra Comput. **3**(4) (1993), 365–409.

[315] A. Yu. Olshanskii, *SQ-universality of hyperbolic groups*, Mat. Sb. **186**(8) (1995), 119–132.

[316] A. Olshanskii and V. Shpilrain, *Linear average-case complexity of algorithmic problems in groups*, J. Algebra **668** (2025), 390–419.

[317] A. Yu. Ol'shanskiĭ, *Almost every group is hyperbolic*, Int. J. Algebra Comput. **2** (1992), 1–17.

[318] S. Orevkov, *Automorphism group of the commutator subgroup of the braid group*, Ann. Fac. Sci. Toulouse **26** (2017), 1137–1161.

[319] D. V. Osin, *The entropy of solvable groups*, Ergod. Theory Dyn. Syst. **23** (2003), 907–918.

[320] D. Osin, *Small cancellations over relatively hyperbolic groups and embedding theorems*, Ann. Math. **172** (2010), 1–39.

[321] P. Papasoglu, An algorithm detecting hyperbolicity, Geometric and computational perspectives on infinite groups, in: *DIMACS Ser. Discrete Math. Theoret. Comput. Sci.*, **25**, 193–200, Amer. Math. Soc., Providence, RI, 1996.

[322] A. I. Papistas, *A note on fixed points of certain relatively free nilpotent groups*, Commun. Algebra **29** (2001), 4693–4699.

[323] D. S. Passman, *The Algebraic Structure of Group Rings*, John Wiley and Sons, New York, 1977.

[324] M. G. Peretjat'kin, *Example of an ω_1-categorical complete finitely axiomatizable theory*, Algebra Log. **19**(3) (1980), 202–229.

[325] A. Pietrowski, *The isomorphism problem for one-relator groups with non-trivial centre*, Math. Z. **136** (1974), 95–106.

[326] E. L. Post, *A variant of a recursively unsolvable problem*, Bull. Am. Math. Soc. **52** (1946), 264–268.

[327] S. Pride, *The isomorphism problem for two-generator one-relator groups with torsion is solvable*, Trans. Am. Math. Soc. **227** (1977), 109–139.
[328] D. Puder, *Primitive words, free factors and measure preservation*, Isr. J. Math. **201** (2014), 25–73.
[329] D. Puder and O. Parzanchevski, *Measure preserving words are primitive*, J. Am. Math. Soc. **28** (2015), 63–97.
[330] D. Puder and C. Wu, *Growth of primitive elements in free groups*, J. Lond. Math. Soc. **90** (2014), 89–104.
[331] A. Razborov, *Systems of equations in a free group*, Math. USSR, Izv. **25**(1) (1985), 115–162.
[332] S. Rees, The development of the theory of automatic groups, in: K. Ohshika and A. Papadopoulos A, eds., *The Tradition of Thurston II. Geometry and Groups*, 449–473, Springer, Cham, 2022. ISBN 978-3-030-97559-3.
[333] A. H. Rhemtulla, *Commutators of certain finitely generated soluble groups*, Can. J. Math. **21** (1969), 1160–1164.
[334] E. Rips, *Subgroups of small cancellation groups*, Bull. Lond. Math. Soc. **14** (1982), 45–47.
[335] C. F. Rocca Jr. and E. C. Turner, Test ranks of finitely generated abelian groups, in: *Combinatorial and Geometric Group Theory* (New York, 2000/Hoboken, NJ, 2001), Contemp. Math., **296**, 199–206, Amer. Math. Soc., Providence, RI, 2002.
[336] J. Roe, *Lectures on Coarse Geometry*, University Lecture Series, **31**, American Mathematical Society, Providence, RI, 2003, viii+175 pp. ISBN 0-8218-3332-4.
[337] V. Roman'kov, *Equations in free metabelian groups*, Sib. Math. J. **20** (1979), 469–471.
[338] V. Roman'kov, *On test elements in free solvable groups of rank 2*, Algebra Log. **40** (2001), 192–201 (Russian).
[339] V. A. Roman'kov, *Diophantine questions in the class of finitely generated nilpotent groups*, J. Group Theory **19** (2016), 497–514.
[340] V. A. Roman'kov, *Polycyclic, metabelian or soluble of type $(FP)_\infty$ groups with Boolean algebra of rational sets and biautomatic soluble groups are virtually abelian*, Glasg. Math. J. **60** (2018), 209–218.
[341] V. A. Roman'kov, N. G. Khisamiev and A. A. Konyrkhanova, *Algebraically and verbally closed subgroups and retracts of finitely generated nilpotent groups*, Sib. Math. J. **58** (2017), 536–545.
[342] G. Rosenberger, Minimal generating systems for plane discontinuous groups and an equation in free groups, in: *Groups—Korea 1988* (Pusan, 1988), Lecture Notes in Math., **1398**, 170–186, Springer, Berlin, 1989.
[343] G. Rosenberger, *The isomorphism problem for cyclically pinched one-relator groups*, J. Pure Appl. Algebra **95** (1994), 75–86.
[344] G. Sacerdote, *Elementary properties of free groups*, Trans. Am. Math. Soc. **178** (1973), 127–138.
[345] S. Schleimer, *Polynomial-time word problems*, Comment. Math. Helv. **83** (2008), 741–765.
[346] Z. Sela, *The isomorphism problem for hyperbolic groups*, Ann. Math. (2) **141** (1995), 217–283.
[347] Z. Sela, *Diophantine geometry over groups. I. Makanin–Razborov diagrams*, Publ. Math. IHÉS **93** (2001), 31–105.
[348] Z. Sela, *Diophantine geometry over groups. II. Completions, closures and formal solutions*, Isr. J. Math. **134** (2003), 173–254.
[349] Z. Sela, *Diophantine geometry over groups. IV. An iterative procedure for validation of a sentence*, Isr. J. Math. **143** (2004), 1–130.
[350] Z. Sela, *Diophantine geometry over groups. III. Rigid and solid solutions*, Isr. J. Math. **147** (2005), 1–73.
[351] Z. Sela, *Diophantine geometry over groups. V1. Quantifier elimination. I*, Isr. J. Math. **150** (2005), 1–197.
[352] Z. Sela, *Diophantine geometry over groups. V2. Quantifier elimination. II*, Geom. Funct. Anal. **16** (2006), 537–706.
[353] Z. Sela, *Diophantine geometry over groups. VI. The elementary theory of a free group*, Geom. Funct. Anal. **16** (2006), 707–730.
[354] Z. Sela, *Diophantine geometry over groups. X. The elementary theory of free products of groups*, preprint, arXiv:1012.0044.
[355] R. Sharp, *Local limit theorems for free groups*, Math. Ann. **321** (2001), 889–904.

[356] V. Shpilrain, *Automorphisms of F/R' groups*, Int. J. Algebra Comput. **1** (1991), 177–184.
[357] V. Shpilrain, *Recognizing automorphisms of the free groups*, Arch. Math. **62** (1994), 385–392.
[358] V. Shpilrain, *Non-commutative determinants and automorphisms of groups*, Commun. Algebra **25** (1997), 559–574.
[359] V. Shpilrain, *Generalized primitive elements of a free group*, Arch. Math. **71** (1998), 270–278.
[360] V. Shpilrain, *Fixed points of endomorphisms of a free metabelian group*, Math. Proc. Camb. Philos. Soc. **123** (1998), 77–85.
[361] V. Shpilrain, *Counting primitive elements of a free group*, Contemp. Math. **372** (2005), 91–97.
[362] V. Shpilrain, *Average-case complexity of the Whitehead problem for free groups*, Commun. Algebra **51** (2023), 799–806.
[363] P. V. Silva and P. Weil, *Automorphic orbits in free groups: words versus subgroups*, Int. J. Algebra Comput. **20** (2010), 561–590.
[364] I. Snopce, S. Tanushevski and P. Zalesskii, *Retracts of free groups and a question of Bergman*, Int. Math. Res. Not. **11** (2022), 8280–8294.
[365] J. Stallings, *Topology of finite graphs*, Invent. Math. **71** (1983), 551–565.
[366] J. Stallings, *Finiteness properties of matrix representations*, Ann. Math. (2) **124** (1986), 337–346.
[367] A. Steinberg, *On roots of $a^k b^l$*, Math. Z. **192** (1986), 1–8.
[368] J. Tao, *Linearly bounded conjugator property for mapping class groups*, Geom. Funct. Anal. **23**(1) (2013), 415–466.
[369] G. Tardos, *On the intersection of subgroups of a free group*, Invent. Math. **108** (1992), 29–36.
[370] G. Tardos, *Towards the Hanna Neumann conjecture using Dicks' method*, Invent. Math. **123** (1996), 95–104.
[371] *The Kourovka Notebook. Unsolved Problems in Group Theory*, Nineteenth edition. Edited by E. I. Khukhro and V. D. Mazurov. Sobolev Institute of Mathematics, Russian Academy of Sciences, Siberian Branch, Novosibirsk, 2018.
[372] E. I. Timoshenko, *Certain algorithmic questions for metabelian groups*, Algebra Log. **12** (1973), 132–137.
[373] E. I. Timoshenko, *Center of a group with one defining relation in the variety of 2-solvable groups*, Sib. Mat. Zh. **14** (1973), 1351–1355, 1368 (Russian).
[374] E. I. Timoshenko, *Test elements and test rank of a free metabelian group*, Sib. Math. J. **41** (2000), 1200–1204.
[375] E. I. Timoshenko, *Computing test rank for a free solvable group*, Algebra Log. **45** (2006), 254–260.
[376] N. Touikan, *A fast algorithm for Stallings' folding process*, Int. J. Algebra Comput. **16** (2006), 1031–1045.
[377] E. Ventura, *On fixed subgroups of maximal rank*, Commun. Algebra **25** (1997), 3361–3375.
[378] S. Wenger, *Nilpotent groups without exactly polynomial Dehn function*, J. Topol. **4** (2011), 141–160.
[379] J. H. C. Whitehead, *On equivalent sets of elements in free groups*, Ann. Math. **37** (1936), 782–800.
[380] J. H. C. Whitehead, *On certain sets of elements in a free group*, Proc. Lond. Math. Soc. **41** (1936), 48–56.
[381] W. Whitten, *Knot complements and groups*, Topology **26** (1987), 41–44.
[382] J. S. Wilson, *On exponential growth and uniformly exponential growth for groups*, Invent. Math. **155** (2004), 287–303.
[383] D. Wise, *A continually descending endomorphism of a finitely generated residually finite group*, Bull. Lond. Math. Soc. **31** (1999), 45–49.
[384] D. Wise, *The residual finiteness of positive one-relator groups*, Comment. Math. Helv. **76** (2001), 314–338.
[385] D. T. Wise, *Cubulating small cancellation groups*, Geom. Funct. Anal. **14** (2004), 150–214.
[386] D. Wise, *Research announcement: the structure of groups with a quasiconvex hierarchy*, Electron. Res. Announc. Math. Sci. **16** (2009), 44–55.
[387] D. Wise, *The Structure of Groups with a Quasiconvex Hierarchy*, Ann. of Math. Stud., **209**, Princeton University Press, Princeton, NJ, 2021, x+357 pp.
[388] D. T. Wise, *Coherence, local indicability and nonpositive immersions*, J. Inst. Math. Jussieu **21**(2) (2022), 659–674.
[389] P. Wright, *Group presentations and formal deformations*, Trans. Am. Math. Soc. **208** (1975), 161–169.

www.ingramcontent.com/pod-product-compliance
Lightning Source LLC
Chambersburg PA
CBHW082353220526
45470CB00008B/2737